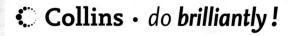

Collins · *do brilliantly !*

Instant**Facts**

Science

A-Z of **essential facts** and definitions

D McMonagle BSc PhD Cchem FRSC

William Collins' dream of knowledge for all began with the publication of his first book in 1819. A self-educated mill worker, he not only enriched millions of lives, but also founded a flourishing publishing house. Today, staying true to this spirit, Collins books are packed with inspiration, innovation and practical expertise. They place you at the centre of a world of possibility and give you exactly what you need to explore it.

Collins. Do more.

Published by Collins
An imprint of HarperCollins*Publishers*
77–85 Fulham Palace Road
Hammersmith
London
W6 8JB

Browse the complete Collins catalogue at

www.collinseducation.com
© HarperCollins*Publishers* Limited 2005

First published as Collins Gem Basic Facts Science 1993

10 9 8 7 6 5 4 3 2 1

ISBN 0 00 720555 4

British Library Cataloguing in Publication Data
A catalogue record for this publication is available from the British Library

Every effort has been made to contact the holders if copyright material, but if any have been inadvertently overlooked, the Publishers will be pleased to make the necessary arrangements at the first opportunity.

Edited and Project Managed by Marie Insall
Production by Katie Butler
Design by Sally Boothroyd/Jerry Fowler
Printed and bound by Printing Express, Hong Kong

You might also like to visit
www.harpercollins.co.uk
The book lover's website

Introduction

Instant Facts Science is one of a series of illustrated A–Z subject reference guides of the key terms and concepts used in the most important school subjects. With its alphabetical arrangement, the book is designed for quick reference to explain the meaning of words used in the subject and so is an excellent companion both to course work and during revision.

Bold words in an entry identify key terms which are explained in greater detail in entries of their own; important terms that do not have separate entries are shown in *italic* and are explained in the entry in which they occur.

Other titles in the *Instant Facts* series include:
English
Modern World History
Biology
Physics
Geography
Maths
Chemistry

A

absorption **1.** The taking in of radiated energy. When **radiation** travelling in one material meets the surface of a second, three things may happen to it:

(a) It may bounce back into the first material: this is called **reflection**.

(b) It may travel on in a new direction through the second material: this is called **refraction**.

(c) It may disappear into the second material: this is called **absorption**.

When a surface absorbs radiation the **energy** must appear in a new form. In most cases this is **heat** energy: the **temperature** increases.

2. (of food) The process by which digested food particles pass from the **alimentary canal** into the bloodstream. In mammals absorption takes place in the ileum, which is part of the small intestine.

absorption

acceleration (*a*) The change of **velocity**, *v*, of an object in unit time. The **SI unit** of acceleration is the metre per second per second, m/s². Acceleration can be calculated using the following equation:

$$a = \frac{v-u}{t}$$

where *u* is the initial velocity and *v* is the final velocity. Acceleration is a vector quantity. An object accelerates if its **speed** and/or its direction of motion change.

When a net outside force, *F*, acts on an object of mass, *m*, the resulting acceleration, *a*, is given by:

$$a = F/m$$

At any moment on a velocity/time graph the acceleration is given by the slope of the graph. Positive acceleration results in an increase in velocity, while negative acceleration (often called *deceleration* or *retardation*) results in a decrease in velocity.

acid A substance which releases **hydrogen ions** (H^+) when added to
water. Acid solutions have a **pH** of less than 7. Common laboratory
acids are:
(a) Strong acids: – nitric acid (HNO_3), **hydrochloric acid** (HCl), **sulphuric
    acid** (H_2SO_4).
(b) Weak acids: – ethanoic acid (CH_3COOH), citric acid ($C_6H_8O_7$).
   Strong acids ionize completely in water, weak acids only partially. A few
acids are corrosive liquids and must be handled with care.
Acids:
(a) Turn blue **litmus** red.
(b) Give **carbon dioxide** when added to carbonates.
(c) Give **hydrogen** when added to certain **metals**.
(d) Neutralize **alkalis**.

acid rain Rain polluted by acids in the atmosphere. Coal often contains
substances which produce the acidic gases sulphur dioxide and nitrogen
oxides when it is burnt. If these acidic gases are allowed to escape into the
atmosphere they dissolve in any moisture present to produce acids. When
the moisture falls to the earth as rain it is, in fact, acid(ic) rain.
   Acid rain is responsible for considerable damage to the environment.
Once soils become highly acidic, plants and animals are adversely affected
and potentially harmful **ions** such as **aluminium** are able to dissolve in the
acidic solution. The acidic solutions may accumulate in lakes where their
effects on wildlife are devastating. Recently steps have been taken to
reduce the quantity of acidic gases produced by British coal-fired power
stations and car exhausts. In future it is planned to fit coal-fired power
stations with flue desulphurization units to remove acidic gases before flue
gases are released into the atmosphere.

active transport The movement of substances against a concentration
gradient (*see* **osmosis**) using energy produced by metabolic processes.
Examples of active transport are (a) the reabsorption of useful substances,
including **glucose** and **amino acids**, in the mammalian **kidney**, and (b) the
uptake of **mineral salts** by the **root** of plant.

air The **mixture** of **gases** which
make up the **atmosphere**
surrounding the **Earth**. The exact
composition of the air varies
slightly from place to place and
with height above the ground.

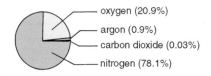

oxygen (20.9%)
argon (0.9%)
carbon dioxide (0.03%)
nitrogen (78.1%)

air *The average composition of pure air.*

Air, particularly in industrial areas, often contains pollutants. Some common examples are shown below:

Pollutant	Sources
Sulphur dioxide	burning coal and oil
Carbon monoxide	engines, cigarettes
Nitrogen oxides	burning coal, car engines
Soot	engines, fires
Pollen	frees, plants
Chlorofluorocarbons (CFCs)	Aerosols

Air is vital for life. The **oxygen** it contains is necessary for **respiration**, and the **carbon dioxide** for **photosynthesis**. *See* **ozone layer**.

alimentary canal

The digestive canal of an animal. In humans this canal is a tube about nine metres long running from the mouth to the anus.

The inner surface of the intestine, and particularly the ileum, consists of finger-like projections called *villi*. Each *villus* has a thin wall and contains a network of blood capillaries. The villi provide a large surface area that allows the efficient absorption of soluble foods.

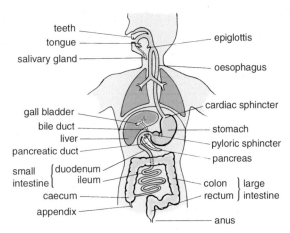

alimentary canal *Diagram showing the canal and digestive organs.*

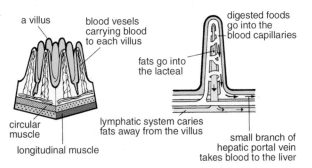

ileum *Section through the ileum wall (left) and a section of a villus (right).*

alkali A **base** which is soluble in **water**. Alkalis are usually metal hydroxides, e.g. sodium hydroxide. **Ammonia** solution is also an alkali. Alkalis are corrosive and should be handled with care. Typical properties of alkalis are:
(a) They turn red **litmus** blue.
(b) They neutralize **acids**.
(c) They have a **pH** of more than 7.
(d) They react with acids to produce a **salt** and water only.

alloy A **mixture** which is made up of two or more **metals**, or which contains metals and **nonmetals**.

The properties of an alloy are different from the sum of the properties of the substances it contains. Alloys are used much more than pure metals. This is because an alloy can be made with a particular set of properties by bringing two or more **elements** together in the right proportions. **Aluminium** is a soft metal, but mixing it with a small amount of **copper** produces the alloy *duralumin* which is strong enough to be used in aircraft frames. Here are some other examples of alloys in common use:

Alloy	Elements it contains
brass	copper and zinc
bronze	copper and tin
pewter	lead, tin and small amounts of antimony
solder	lead and tin
steel	**iron** and **carbon**

alpha particle (α) This consists of two **neutrons** and two **protons** bonded together and thus it carries a positive **charge**. It contains the same particles as the **nucleus** of a helium-4 **atom**. Many **radioactive** nuclei give off alpha particles when they decay. An example is the common **isotope** of **uranium**:

$$^{238}_{92}\text{U} \rightarrow\ ^{234}_{90}\text{Th} + ^{4}_{2}\alpha + \text{energy}$$

The flow of alpha particles from a radioactive source is called alpha **radiation**. The energy transferred appears as the **kinetic energy** of the alpha particles. Subsequently, alpha particles collide with other particles and slow down: there is energy transfer to the substance giving a small temperature rise. As alpha radiation loses energy in matter, it causes ions to appear, hence it is an **ionizing radiation**.

Matter slows down alpha radiation very quickly. It has a range of only a few centimetres in air and cannot pass through even thin card.

alternating current (AC) A **current** which flows alternately in one direction, then in the opposite direction, around a circuit.

The **frequency** of the alternating current mains supply in the United Kingdom is 50 Hz (cycles per second).

aluminium The most abundant **metal** in the earth's crust. Aluminium has **valency** 3 and is very reactive. It is obtained by the **electrolysis** of a molten mixture of aluminium oxide, dissolved in another mineral called cryolite, using graphite **electrodes** in a *Hall-Hérault cell*.

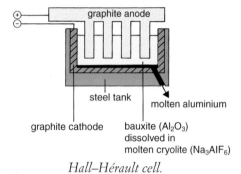

Hall–Hérault cell.

The reactions at the electrodes are:

Anode $\quad 2O^{2-} - 4e^- \rightarrow O_2$

Cathode $\quad Al^{3+} + 3e^- \rightarrow Al$

Due to the high temperature at which this process operates, the graphite anodes react with the oxygen form to give carbon dioxide and must be replaced at regular intervals.

Aluminium and its **alloys** are resistant to corrosion because of a protective layer of oxide which forms on the surface of the metal. (This protective layer can be thickened by a process called *anodizing*.)

amino acids *Organic compounds* which are sub-units of **proteins**. About seventy different amino acids are known but only twenty to twenty-four are found in living organisms.

Amino acids are bonded together in chains known as *peptides*. The link between adjacent amino acids is called a peptide bond or link.

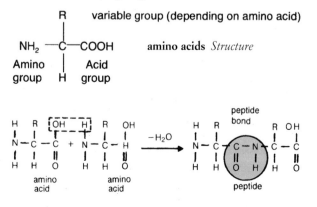

amino acids Structure

amino acids
A peptide bond.

When many amino acids are joined together in this way the whole complex is called a *polypeptide* and this is the basis of protein structure.

ammonia A colourless **gas** with an unpleasant odour. It is very soluble in water and gives an alkaline solution (*see* **alkali**) which is sometimes called ammonium hydroxide. Ammonia is a **covalent compound**. It has a characteristic reaction with the gas hydrogen chloride, producing dense white fumes of ammonium chloride:

$$NH_3(g) + HCl(g) \rightarrow NH_4Cl(s)$$

Most ammonia is made by the *Haber process*. Nitrogen and **hydrogen** in the ratio of 1:3 react at 500 °C and 20 MPa pressure on an iron **catalyst**:

$$N_2(g) + 3H_2(g) \rightarrow 2NH_3(g)$$

Ammonia is produced by **bacteria** found on the roots of *leguminous* plants like peas and beans. Also, when **proteins** decompose, ammonia is released. Both of these are important sources of plant nutrients (*see* **nitrogen cycle**).

anhydrous Containing no **water**. The term is usually used to describe **salts** with no *water of crystallization*. For example:

	Anhydrous salt	**Hydrated salt**
Copper(II) sulphate	$CuSO_4$	$CuSO_4.5H_2O$
Sodium carbonate	Na_2CO_3	$Na_2CO_3.10H_2O$

The term may also be used to describe **liquids** that are perfectly dry and contain no water, e.g. anhydrous ether.

antibiotics Substances formed by certain **bacteria** and **fungi** which inhibit the growth of other microorganisms. Common examples are penicillin and streptomycin. Many are now made synthetically.

antibodies **Proteins** produced by the **tissues** of vertebrates in response to **antigens**. Antibodies react with antigens to make them harmless.

antigens   antibodies   antigens neutralized

antibodies *Antigens are deactivated by antibodies.*

antigens Materials which are foreign to an organism. Such materials include microorganisms such as **bacteria**, and their toxins, or transplanted **organs** or **tissue**. *See* **antibodies**.

artery A blood vessel which transports **blood** away from the **heart** to the **tissues**. In mammals, arteries (except for the pulmonary artery) carry

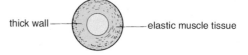

thick wall — elastic muscle tissue

artery *Section through an artery.*

bright red oxygenated blood. They divide into smaller vessels called *arterioles* which themselves eventually sub-divide into **capillaries**. Arteries have thick, elastic, muscular walls as they need to withstand the high **pressure** caused by the **heartbeat**. Compare **vein**.

asexual reproduction Reproduction in which new organisms are formed from a single parent. Asexual reproduction does not involve the production of **gametes**. The offspring are genetically identical to each other and to the parent organism. They are sometimes referred to as clones. *See* **vegetative reproduction**.

assimilation The process by which food that has already been digested is incorporated into or stored in the cells of the organism. **Amino acids**, for example, are built up into **protein**, which is used for cell growth or maintenance. In mammals, any **glucose** which is not required immediately to provide energy in tissue **respiration** is converted in **liver** and **muscle** cells to **glycogen**. This short-term energy store can be reconverted to glucose if the blood glucose level falls (*see* **insulin**). Excess glucose not stored as glycogen is converted into **fat** and stored in fat storage cells beneath the skin. Fat is a long-term energy store. Excess amino acids cannot be stored and are broken down in the liver.

atmosphere
1. A unit of **pressure** equal to 101 325 pascals. Atmospheric pressure is the result of the **weight**

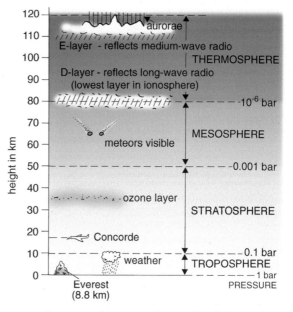

atmosphere *Earth's atmosphere is divided into zones.*

of air pushing downwards. It varies from place to place and time to time but is always about 1 atmosphere near sea level. Atmospheric pressure decreases as height above the ground increases.

2. The **mixture** of **gases** which surrounds a **planet**. The atmosphere on **Earth** is **air**. The Earth's atmosphere is often divided into several regions or zones. (*See* diagram on page 7).

atom The smallest particle of an **element** that can exist. Atoms are the building blocks of which everything is made. They are made up of even smaller *subatomic particles*.

Subatomic particle	Position in atom	Electric charge	Relative mass
Proton	(at the centre)	positive	1
Neutron	(in the **nucleus**)	neutral	1
Electron	moving around the nucleus	negative	1/1840

The proton and electron carry equal but opposite charges. The atom as a whole is neutral, hence the number of protons always equals the number of electrons. All atoms of the same element have the same number of protons and hence the same **atomic number**, but atoms of the same element may have different numbers of neutrons. *See* **isotopes**.

Atoms are the smallest part of an element that can take part in a chemical reaction.

atomic number (Z) The number of **protons** in the **nucleus** of an **atom**. All atoms of the same **element** have the same atomic number. An atom contains the same number of **electrons** as protons, so the number of electrons also equals the atomic number.

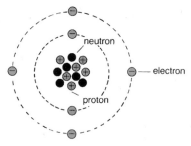

atom *An atom of carbon.*

B

backbone *See* **vertebral column**.

background radiation The radiation, from natural sources, which is always present in the earth's atmosphere. Radioactive matter exists in almost all of the rocks of the earth's crust. As it **decays** it gives off **radiation**. *Cosmic radiation* is also continually hitting the earth from space. As a result of the above two processes a radiation detector, such as a *Geiger-Müller tube*, gives a count of about one radiation a second. Background radiation has to be taken into account when measuring the activity of other radioactive sources. The activities on earth which produce radiation, such as nuclear weapons, nuclear power stations and the medical applications of radioactive **isotopes**, appear to have little if any effect on the amount of background radiation.

bacteria Organisms consisting of only one cell of diameter 1–2 microns. Some bacteria are harmful because they cause diseases such as *tetanus*. Some bacteria are useful, for example, as sources of **antibiotics**.

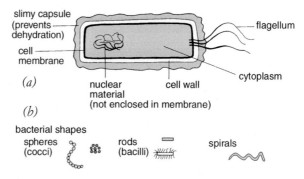

bacteria (*a*) *Generalised structure of a bacterium.* (*b*) *bacterial shapes.*

balanced diet
A diet which contains the correct nutritional components needed for the organism to remain healthy. The term is usually used in connection with humans and domestic animals. A balanced diet for humans should contain the following:
(a) **Protein**
(b) **Carbohydrate**
(c) **Fat**
(d) **Roughage**
(e) **Vitamins**
(f) **Water**
(g) **Mineral salts**

   The amounts of carbohydrate and fat should be just enough to supply the body with its energy requirements. Excess carbohydrates and fats are undesirable as they increase the amount of fat stored in the body.

base A substance which reacts with an **acid** to form a **salt** and **water** only. Bases are usually *metal oxides* or *hydroxides*, e.g. sodium hydroxide (NaOH), copper(II) oxide (CuO). Metal oxides and hydroxides which are soluble in water are known as **alkalis**. These are usually compounds of **group I** or **II metals**.

battery A group of two or more single electric cells connected together in series. A car battery consists of six 2V cells giving a total voltage of 12V.

bell A device constructed to make a ringing sound. An electric bell is a device based on an **electromagnet**. When the switch is closed the circuit is completed.The iron core becomes a magnet and attracts the soft iron armature. As the armature moves across, striking the bell on the left, the circuit is

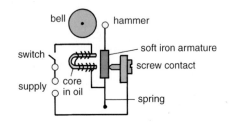

bell *Circuit diagram of an electric bell.*

broken. The spring returns the armature to its original position causing it to strike the bell on the right. This cycle is continually repeated so long as the switch is closed. Buzzers work on a similar principle.

beta particle (β) A high-speed **electron** which carries negative **charge**. It is emitted by a **nucleus** during radioactive decay or nuclear **fission**. The energy transferred appears as the **kinetic energy** of the beta particle. A flow of beta particles from a radioactive source is called beta **radiation**. Beta radiation causes ions to appear when it hits matter, hence it is an **ionizing radiation**. Beta radiation travels further into matter than alpha radiation (*see* **alpha particles**) but not as far as **gamma radiation**. It is completely absorbed by a few millimetres of aluminium.

bile A green alkaline fluid which is produced in the **liver** of mammals. Bile is stored in the gall bladder and passes into the duodenum through the bile duct. In the duodenum it causes fats to be broken into tiny droplets prior to **digestion**. This process is called *emulsification* and it speeds up fat digestion.

bimetallic strip A device made of strips of two different **metals** (hence bimetallic) fixed together.For a given **temperature** change one metal expands (*see* **expansion**) more than the other. This causes the bimetallic strip to bend. The amount which it bends depends on the size of the

temperature change.
Bimetallic strips are used in
simple **thermometers** and
thermostats.

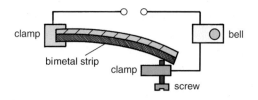

When the temperature
reaches a certain value,
determined by the setting of
the screw, the bimetallic
strip bends sufficiently to
complete the circuit and a
bell starts to ring.

bimetallic strip *One type used in a simple fire alarm.*

biodegradable Capable of being broken down by biological action.
This action usually involves **bacteria** and **fungi**. Biodegradable waste
products are usually easy to dispose of and are not a source of **pollution**.
Nonbiodegradable materials, such as plastics, are not broken down by the
action of bacteria and fungi. They are difficult to dispose of and are often
a source of pollution.

biosphere The parts of the **Earth** and the **atmosphere** collectively which
are inhabited by living things. Living things are found:
(a) In the atmosphere immediately above the earth.
(b) On the surface of the earth.
(c) Within the surface of the earth.
(d) In lakes, rivers and oceans.
   Many scientists believe that the proper functioning of the biosphere is
under threat from **pollution**.

birth The human baby is born as a result of muscular contractions of the
uterus wall. The amniotic fluid escapes and the baby is pushed through the
cervix and vagina, thus leaving the mother's body.
   After the baby is born the umbilical cord, which attached it to the
placenta, is cut. The placenta and the remaining part of the umbilical cord
are then expelled from the uterus as afterbirth. The part of the umbilical
cord attached to the baby will drop off within a few days. After birth the
baby must use its own **lungs** for **gas exchange** and rely on its own digestive
system for food.

bitumen A black tarry **mixture** of high **boiling point hydrocarbons** which
is left behind in the distillation of **petroleum**. It is impermeable to water
and is widely used for roofing and, mixed with sand, for road surfacing.

blast furnace A furnace which allows the continuous production of molten **iron**. The furnace is charged with a mixture of coke, iron ore and limestone. At the bottom of the furnace coke and **air (oxygen)** react to produce **carbon dioxide**. This is an **exothermic reaction** and the **temperature** rises to 1800 °C:

$$C(s) + O_2(g) \rightarrow CO_2(g)$$

More coke then reacts with the carbon dioxide to produce carbon monoxide:

$$C(s) + CO_2(g) \rightarrow 2CO(g)$$

The carbon monoxide reduces the iron oxides to iron. For example:

$$Fe_2O_3(s) + 3CO(g) \rightarrow 3CO_2(g) + 2Fe(l)$$

The molten iron flows to the bottom of the furnace where it is tapped off.
   The limestone in the charge is decomposed by the heat, producing calcium oxide and carbon dioxide. The calcium oxide reacts with impurities from the iron ore (mainly silica SiO_2) and forms a molten slag:

$$CaO(s) + SiO_2(s) \rightarrow CaSiO_3(l)$$

   The slag also sinks to the bottom of the furnace where it floats on the molten iron. The iron produced by a blast furnace contains about 3% carbon and is very brittle. In **steel** making the amount of carbon and other impurities is reduced by blowing oxygen into the molten iron.

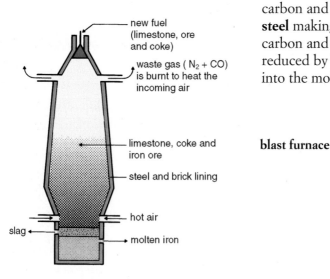

new fuel (limestone, ore and coke)

waste gas (N_2 + CO) is burnt to heat the incoming air

limestone, coke and iron ore

steel and brick lining

hot air

slag ←

molten iron

blast furnace

blood A fluid **tissue** which is found in many animals. The principal function of blood is to transport substances from one part of the body to another. The blood of mammals consists of a watery solution called **plasma** in which there are three types of **cells**: *platelets*, **red blood cells** and **white blood cells**.

The main functions of the blood in humans are:
(a) Transport of **oxygen** from the **lungs** to the tissues.
(b) Transport of **carbon dioxide** from the tissues to the lungs.
(c) Transport of toxic by-products to the **organs** of **excretion**.
(d) Transport of **hormones** from **endocrine glands** to target organs.
(e) Transport of digested food from the ileum to the tissues.
(f) Prevention of infection by:
   (i) **blood clotting**
   (ii) *phagocytosis* by white blood cells
   (iii) **antibody** production.

blood clotting
A thickening of the blood which occurs when blood is exposed to the **air** as a result of injury. The

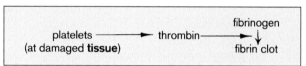

blood platelets produce an **enzyme** (thrombin) which causes the conversion of soluble plasma protein (fibrinogen) into fibrin. The fibrin forms a meshwork of fibres into which platelets become lodged. The resulting clot plugs the wound, restricting blood loss and the entry of microorganisms.

blood clotting *The process of clot formation.*

boiling point The **temperature** at which a **liquid** boils to become a **gas**. It is not fixed but depends on atmospheric **pressure**. The higher the atmospheric pressure, the higher the boiling point. Impurities in the liquid also cause the boiling point to rise. The **energy** needed to change a liquid at its boiling point into a gas is called **latent heat**.

bond The chemical link which holds **atoms** together in **molecules** and **giant structures**. Bonds are formed between atoms and are generated by **electrons**. Some atoms lose or gain electrons, forming **ions**, giving rise to **ionic bonds**. Other atoms share pairs of electrons, giving rise to **covalent bonds**.

bone The **tissue** of which the vertebrate **skeleton** is made. It contains the **protein** *collagen* which gives tensile strength, and calcium phosphate which gives bone its hardness. Some bones have a hollow cavity containing *bone marrow*. New **red blood cells** are made in the bone marrow.

brain A large mass of nerve **cells** in animals whose function is to coordinate the processes of the body. In vertebrates the brain is found at the head of the body. It is protected by the cranium and connected to the body by the spinal nerves in the **spinal cord** and by cranial nerves such as the optic nerve and the auditory nerve.

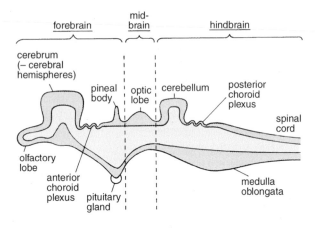

brain *A generalized scheme of the vertebrate brain.*

Within the human brain there are millions of nerve cells which are continually receiving and sending out *nerve impulses*. The brain is able to interpret these electrical impulses in such a way that environmental **stimuli** such as **light** and **sound** are appreciated. This allows the recipient of the stimuli to respond and adapt to the **environment** as appropriate to the situation.

The brain also coordinates all body activity to ensure efficient operation. It stores information (memory) so that behaviour can be modified as the result of past experience.

breathing (in mammals) The process of inhaling and exhaling **air** for the purpose of **gas exchange**. In mammals the gas exchange surface is situated in the **lungs**.

The exchange of air in the lungs (ventilation) is caused by changes in the **volume** of the thorax. This change in volume is

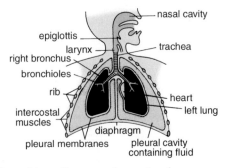

breathing *Lungs and associated structures.*

achieved by the action of
the diaphragm and the
muscles between the ribs
(inter-costal muscles).

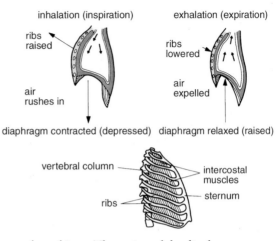

When the diaphragm
contracts it depresses, and
this increases the volume
of the thorax. The air
pressure outside the
lungs (atmospheric
pressure) is now greater
than the air pressure
inside, hence air is forced
into the lungs. When the
diaphragm is relaxed it

breathing *The action of the diaphragm.*

reduces the volume of the thorax. Air pressure in the lungs is now greater
than outside, and air is forced out of the lungs.

The action of the diaphragm is accompanied by the raising and lowering
of the rib cage to accommodate the changes in the volume of the lungs.
These rib cage movements are the result of contraction and relaxation of
the intercostal muscles.

The rate of lung ventilation is called the *breathing rate*. The breathing
rate in humans is controlled by a part of the **brain** called the medulla
oblongata (*see* **heartbeat**). Increased activity causes an increase in the
breathing rate.

C

calcium A soft grey metallic **element** from Group II of the **Periodic Table**. It is a fairly reactive **metal** and produces a steady stream of **hydrogen** when added to cold water. Calcium compounds are found in many common rocks such as limestone and chalk ($CaCO_3$).

Calcium metal is obtained by the **electrolysis** of molten calcium chloride.

Calcium is essential to the health of plants and animals. Calcium compounds are used by plants in building cell walls and in animals for building **bones**, and **teeth** (*see* **mineral salts**).

Calcium compounds figure prominently in our everyday lives:

(a) *Calcium carbonate* ($CaCO_3$). As limestone this is an important building stone and is used in the production of cement. Lime (calcium oxide) is produced when calcium carbonate is heated in a kiln. Limestone is also used in the production of steel.

(b) *Calcium chloride* ($CaCl_2$). **Anhydrous** calcium chloride is used as a drying agent.

(c) *Calcium hydrogencarbonate* ($Ca(HCO_3)_2$). This chemical is responsible for temporary hardness of water.

(d) *Calcium hydroxide* ($Ca(OH)_2$). This is slaked lime, made by adding lime to water. The resulting limewater is used in the production of mortar.

(e) *Calcium oxide* (CaO). This is lime. It is used in agriculture to combat excessive acidity in soils.

(f) *Calcium sulphate* ($CaSO_4$). The chemical responsible for permanent hardness in water.

capillaries The smallest blood vessels, formed from arterioles and eventually draining into venules and then **veins**. Capillaries form a network within vertebrate **tissue**. The capillary walls are only one cell thick and allow the **diffusion** of substances between the blood and the tissues via a liquid called tissue fluid (*lymph*).

carbohydrates These are *organic compounds* which contain the elements

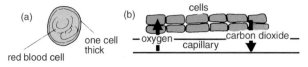

(a) red blood cell — one cell thick

(b) cells — oxygen — capillary — carbon dioxide

capillaries (*a*) *Section through a capillary.* (*b*) *Capillaries and tissues.*

carbon (C), **hydrogen** (H) and **oxygen** (O) and have the general formula $(CH_2O)n$.

Carbohydrates are found either as single **sugar** units or as chains of two or more sugar units bonded together. There are three main groups of carbohydrates; *monosaccharides*, *disaccharides* and *polysaccharides*. Simple carbohydrates, particularly **glucose**, are the **energy** source within living **cells**. 1 g of carbohydrate contains 17.1 kJ of energy.

Long-chain carbohydrates form some structural parts of cells, for example, **cellulose** in plant cell walls. They also act as food reserves, for example, **glycogen** in animals and **starch** in plants.

carbon A nonmetallic **element** which is in Group IV of the **Periodic Table**. It is found in nature in two different forms or *allotropes*; diamond and graphite. All living tissue contains carbon compounds (*organic compounds*), e.g. **carbohydrates, proteins, fats**, and life would not be possible without these.

Carbon burns in **air** or **oxygen** to produce **heat** and **light** energy. Coke, **coal** and charcoal are all impure forms of carbon. They are often used as **fuels**.

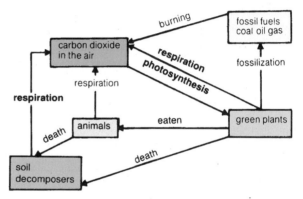

carbon cycle *A summary of the cycle.*

carbon cycle The circulation of the **element carbon** and its **compounds** in nature. Most of this cycle is the result of the metabolic processes of living organisms.

carbon dioxide A gas which makes up 0.03% of **air**. It is formed by a variety of processes:

(a) By the **combustion** of **carbon** in a plentiful supply of **air** or **oxygen**. (Combustion in a limited supply of air will also produce the poisonous gas carbon monoxide).

(b) During the **fermentation** of **sugars**.

(c) As a waste product of **respiration**.

(d) By the action of **heat** (in most cases) or acids on carbonates in the laboratory.

Carbon dioxide plays a vital role in both **photosynthesis** and the **carbon cycle**.

Carbon dioxide turns limewater milky or it turns a bicarbonate indicator yellow. The amount of carbon dioxide in the atmosphere has been steadily increasing over this century. Some scientists believe this is responsible for a small increase in the **temperature** on the surface of the earth. *See* **greenhouse effect**.

carpel The female part of a **flower**. It contains an **ovary** in which there is a varying number of ovules containing embryo sacs. Within the embryo sacs are the female **gametes**.

catalyst A substance which alters the rate of a chemical reaction either speeding it up or slowing it down. The catalyst remains unused at the end of the reaction. The process is called *catalysis*. The transition metals or their **compounds** are often useful catalysts. Some examples are shown below:

Reaction	Catalyst
Haber process	Iron
Contact process	Vanadium(v) oxide (V_2O_5)
Ammonia→nitric acid	Platinum/rhodium alloy
Hardening fats	Nickel

catalyst *Transition metal catalysts.*

Almost all of the chemical reactions which go on inside animals and plants are controlled by catalysts. These organic catalysts are called **enzymes**.

cell   **1.** A unit of cytoplasm controlled by a single **nucleus** and surrounded by a selectively permeable membrane. Cells are the basic units of which most living things are made.

2. A device which provides a transfer to electrical **energy**. For example, a dry cell (commonly but incorrectly called a **battery**) such as might be used

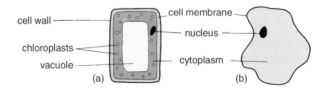

cell *Structure of (a) a plant cell and (b) an animal cell.*

in a torch, gives electricity as the chemical potential energy is transferred to electrical energy.

A *photoelectric* cell provides transfer from light to electric energy. Some calculators and watches are fitted with this type of cell.

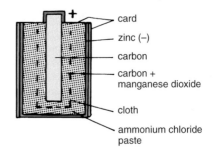

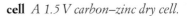

cell *A 1.5 V carbon–zinc dry cell.*

cellulose A **carbohydrate** which forms the framework and gives strength to plant cell walls. Humans are unable to digest cellulose; however, it has an important role as **roughage**. In mammals which are herbivores, populations of **bacteria** are present in the caecum and appendix. They produce an **enzyme** called *cellulase* which digests cellulose into **glucose**.

cement A substance which is used to bind other substances together. Portland cement is widely used in building work. It is made by heating a mixture of limestone and clay to about 1700 °C. The product is ground with gypsum. Cement is mainly composed of calcium silicates and aluminates. When cement powder is mixed with water it undergoes a **chemical change** and sets hard. *Concrete* is formed when cement is mixed with sand or gravel (often called aggregate) and water. The strength of the concrete is determined by the proportion of cement in the mixture.

ceramics Materials made from inorganic chemicals (*see* **inorganic chemistry**) by high temperature processes. Ceramics are resistant to chemical attack and range from semiconductors to electrical insulators. A wide range of materials may be described as ceramics including abrasives, **cement** and cement products, glass and pottery.

chain reaction A situation where one event causes a second which, in turn, causes a third, and so on. An important chain reaction is the one which occurs in a **nuclear power** station. When the **nucleus** of an **atom** of **uranium**-235 absorbs a **neutron** it undergoes nuclear **fission**.

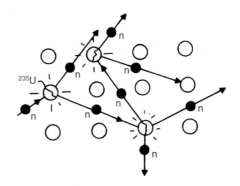

chain reaction *The chain reaction which occurs in a nuclear power station.*

This produces two smaller nuclei, **energy** and three more neutrons. The three neutrons can, in turn, hit the nuclei of three more atoms of uranium-235 producing nine neutrons and so on.

charge A property of some subatomic particles. Two types of charge exist and they have the names *positive* (as on the **proton**) and *negative* (as on the **electron**).

Opposite charges attract each other while like charges repel each other. Normal **atoms** have equal numbers of protons and electrons; they are neutral and have no net charge. When electrons are gained or lost by atoms charged **ions** are produced.

Charge in motion is called **current** and involves **energy** transfer. The unit of charge is the *coulomb* (C). This is the total charge carried by about 1.6×10^{19} electrons – it is the charge transferred by one ampere in one second.

chemical change A change in which one or more chemical substances are changed into different substances by the breaking and making of chemical **bonds** between the **atoms**. Chemical change is usually accompanied by the giving out or taking in of **heat energy**. *See* **exothermic reaction**, **endothermic reaction**. Compare **physical change**.

chlorine (Cl_2) The second of the **halogen** (Group VII) **elements**. It is a green **gas** at room **temperature** and is very reactive. It is very unpleasant if breathed in: it attacks the **lungs** and throat and produces a choking effect. Chlorine has been used as a chemical weapon.

Chlorine occurs naturally as chlorides: sodium chloride is abundant in sea water. It is extracted by the **electrolysis** of sodium chloride solution (during the industrial production of sodium hydroxide). Chlorine is used widely in a range of applications:

(a) It is used in the manufacture of chemicals such as **polymers** (PVC), **pesticides** (DDT), disinfectants (TCP) and solvents (for dry cleaning).

(b) It is added to drinking water and to water in swimming pools in order to kill dangerous **bacteria**. The process is called chlorination.

Chlorine is a vigorous oxidizing agent which readily reacts with most elements. It is prepared in the laboratory by the **oxidation** of concentrated **hydrochloric acid**.

concentrated hydrochloric acid	+	potassium manganate (VII)	$\rightarrow$	chlorine

Metal chlorides. These are usually **ionic compounds**, e.g. sodium chloride (NaCl), barium chloride (BaCl$_2$) (*see* **salts**). They react with concentrated **acids** to produce hydrogen chloride gas.

$$\text{H}_2\text{SO}_4(l) \quad + \quad \text{NaCl(s)} \quad \rightarrow \quad \text{NaHSO}_4(s) \quad + \quad \text{HCl(g)}$$

| concentrated sulphuric acid | sodium chloride | sodium hydrogen-carbonate | hydrogen chloride |

Nonmetal chlorides. These are **covalent compounds**. They are usually either low **boiling point liquids** such as tetrachloromethane (CCl$_4$) or gases such as hydrogen chloride.

chlorophyll A green pigment found in the **chloroplasts** of plant **cells**. This pigment absorbs the light **energy** needed for **photosynthesis**.

chloroplasts Structures found in the cytoplasm of green plant **cells**. Chloroplasts contain the green pigment **chlorophyll** and **photosynthesis** occurs in them.

chromatography An important technique for separating **mixtures** of solutes in a solvent. The following is a simple method of separating the different dyes in a sample of ink.

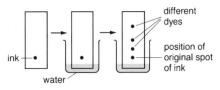

chromatography *Method of separating different dyes in a sample of ink.*

(a) Place a small spot of ink onto a piece of filter paper.
(b) Stand the filter paper in a beaker containing a small amount of water. The level of water must be below the position of the ink spot. Leave the filter paper until water has risen up to the top of it.
(c) The different dyes in the ink will be carried up the filter paper by the water at different rates, hence they will be separated.

This process occurs because some substances cling to the paper more tightly than others. Chromatography has a wide range of applications including the separation of **proteins**, forensic samples, **enzymes** and **viruses**. It works with very small samples.

chromosomes The hereditary material contained within the **nucleus** of a **cell**. The information they contain causes features of one generation to be passed on to the next. Each **species** has characteristic numbers and types of chromosomes.

In humans the *chromosome number* is 46. When a nucleus divides by **mitosis** this *diploid* number (46) of chromosomes is maintained in the new nuclei formed. When a nucleus divides by **meiosis** the new nuclei formed have only half of the diploid number of chromosomes (23) and they are called *haploid* nuclei. Two haploid **gametes** join to form a diploid *zygote*.

In diploid cells the chromosomes occur in similar pairs known as *homologous pairs*. Thus a human diploid cell contains 23 pairs of *homologous chromosomes*.

Chromosomes control the activity of the cell. They are made of many subunits called **genes** which contain coded information in the form of the substance **DNA**.

circuit A number of electrical components joined by *conductors* to a source of electrical **energy**. *Circuit diagrams* show, in simple form, the order in which the components are connected. (*See* Appendix B for symbols used.)

Circuits (a) and (b) show two lamps connected to two cells. In circuit (a) the bulbs are connected in *series* while in circuit (b) they are connected in *parallel*. In circuit (c) an *ammeter* is connected in series with a lamp to measure the **current** in it and a *voltmeter* is connected in parallel to measure the **voltage** across it.

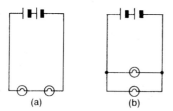

(a)   (b)

circuit (*a*) *A series circuit.* (*b*) *A parallel circuit.*

circuit *Series and parallel circuits combined.*

circulatory system Any system of vessels in animals through which fluid circulates; for example **blood** circulation and the lymphatic system.

In mammals there are two overlapping blood circulations:
(a) Circulation between the **heart** and the **lungs**.
(b) Circulation between the heart and the body.
This arrangement is called a *double circulatory system*. The heart acts as the pump for both circulations and forces blood through the circulatory system always in the same direction.

circulatory system
The double circulatory system of mammals.

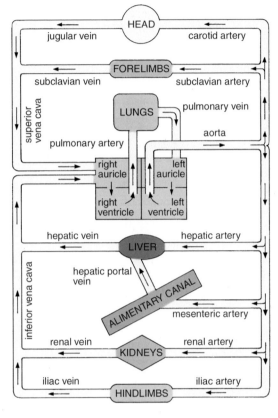

circulatory system *The double circulatory system of mammals.*

classification This is a method of arranging living organisms into groups on the basis of similarities of structure. The present system of classification was devised by Carl von Linné (Linnaeus) in the 18th century. Organisms are first sorted into large groups called *kingdoms* which are divided into smaller groups called *phyla* in animals, and *divisions* in plants. Further subdivisions occur producing subsets contrifling fewer and fewer organisms which have more and more common features.

Ultimately genera (singular genus) are divided into groups of closely related **species**. Some examples are shown in the Table below:

	Human	**Dog**	**Oak**	**Meadow buttercup**
Kingdom	Animal	Animal	Plant	Plant
Phylum/Order	Chordata	Chordata	Spermatophyta	Spermatophyta
Class	Mammalia	Mammalia	Angiospermae	Angiospermae
Order	Primates	Carnivora	Fagales	Ranales
Family	Hominidae	Canidae	Fagaceae	Ranunculaceae
Genus	Homo	Cani	Quercus	Ranunculus
Species	sapiens	familiaris	robur	acris

classification *The scheme of arranging living organisms into subdivisions.*

clone A group of organisms which are *genetically identical*. Clones are the result of **asexual reproduction**.

parent   asexual   identical
organism $\rightarrow$ reproduction $\rightarrow$ offspring (clones)

For example, strawberry runners are clones of a single parent plant.

coal A **fossil fuel** formed over millions of years from fossilized plant material. It has a complex chemical structure and contains **compounds** made of **carbon**, **hydrogen**, **oxygen**, nitrogen and **sulphur**. Coal is used as a **fuel** to produce **heat** in power stations, industry and the home. The compounds containing sulphur and nitrogen produce acidic **gases** when they are burned. These gases are responsible for **acid rain**.

If coal is heated in the absence of air, coal tar is produced. Earlier in this century this was an important source of a whole range of organic chemicals such as phenol which is used in the preparation of dyes, **drugs** and **polymers**. Now most of these products are made from **petroleum**; however, as the world's petroleum reserves decrease, scientists are trying to find waysof converting coal into petroleum products.

About 20% of coal is used to make coke. *See* **resources**.

colour The name we give to our perception of different wavelengths of visible light. The colour of visible **light** varies with wavelength. The wavelength range of visible light is about 400–760 nm. This range contains several hundred *hues* (distinguishable colours) which merge into each other as we go from the deepest red to the deepest violet. The range is often divided into six or seven bands.

λ/nm	Colour
400–420	violet
420–450	indigo
450–500	blue
500–560	green
560–600	yellow
600–650	orange
650–760	red

White light is a mixture of equal amounts of all the different hues. **Refraction** of a parallel beam of white light by a prism can produce the **spectrum** of colours listed above. This is also possible by another method using **diffraction**. The process of separating white light into its constituent colours is called **dispersion**.

Red, blue and green are the three *primary colours*. A mixture of any two of these produces a secondary colour and a mixture of all three produces white light. Mixing coloured light is called *mixing by addition*.

There are few surfaces which reflect all light. Most absorb some wavelengths and reflect others. An object will appear the colour of the light which it reflects. For example, when red paint is illuminated with white light the paint will absorb all colours except red. The red is reflected, making the paint look red. An object which absorbs all colours appears black.

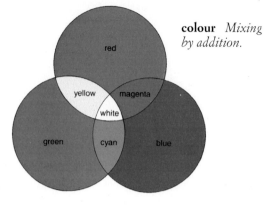

colour *Mixing by addition.*

Most paints contain several different pigments each of which will absorb particular colours. Mixing different coloured paints does not produce the same results as mixing different coloured lights.

combustion In the process of combustion (burning), substances undergo a chemical reaction with **oxygen** (often from the **air**). **Heat** and **light** are usually produced during combustion and oxidation takes place.

community Groups of different **species** of plants and animals living and interacting together within a particular **habitat**. For example, in a small pond there may be different species of fish, amphibians, insect larvae, plants etc, living and interacting together.

compound A pure substance which is made of atoms of two or more **elements** chemically bonded together. The **properties** of compounds are quite different from the properties of the elements from which they are made, e.g. sodium is a poisonous metal which reacts very violently with water, chlorine is a poisonous gas with a choking smell, yet sodium chloride is used in cooking and is essential to life. The atoms in a compound may be held together by either **ionic** or **covalent bonds**: e.g. methane CH_4, water H_2O, sodium chloride NaCl. *See* **mixture**.

concentration The concentration of a *solution* is a measure of how much *solute* (solid) is dissolved in the solution. It is usually expressed in terms of **mass** (e.g. grams), or number of particles (e.g. **moles**) per unit **volume**. For example, the concentration of 1 **dm**^{3} of sodium hydroxide solution containing 40 g of sodium hydroxide (relative molecular mass: NaOH=40) could be written as:

40 g/dm^3 (g/l) or 1 mol/dm^3 (= 1 M)

conduction 1. *Electrical.* The process of transferring a **charge** through a medium. In nearly all cases there is also an **energy** transfer.

Substances can be classified according to the ease with which they allow a charge to pass.

(a) A *conductor* is a substance which is able to pass charge with ease.

(b) A *semiconductor* is fairly good at passing charge.

(c) An *insulator* will only allow charge transfer when the **voltage** applied to it is very high.

Applying a voltage to a sample involves making one end of it positive with respect to the other. If the sample contains charges which are free to move they will be drawn towards the opposite charged end. **Metals** and **carbon** are the only **solids** which will allow charge to pass with ease. This is because a metal contains free **electrons** which are able to move about. The applied voltage makes the electrons accelerate. As they move they collide with each other and with the positive **ions** which form the bulk of the metal. These collisions cause the electrons to slow down; they introduce **resistance** to the flow of charge. There is also energy transfer which causes **temperature** rise (see **fuse**).

The flow of charge per second is called the **current**. It depends on the voltage and on the resistance.

It is normal to give current direction as positive to negative (*conventional current*). However, when the charge carriers are electrons, the current is in fact a flow of negative charge moving from negative to positive (*actual current*).

Apart from liquid metals, other **liquids** will conduct if they contain positive and negative ions. These are compounds which are either molten or in solution and are called *electrolytes*. In this case the flow of charge causes chemical changes; the process is called **electrolysis**. In insulators all the electrons are either firmly attached to the atomic nuclei or involved in chemical bonding; they are not free to move.

2. *Thermal.* The transfer of **energy** from particle to particle in matter. Substances differ in their ability to conduct heat. Good conductors of heat are invariably good conductors of electricity. **Metals** are good thermal conductors.

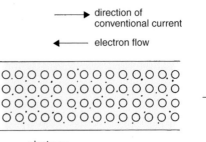

conduction *Electrical conduction.*

They contain free moving **electrons** which carry energy rapidly from points at a high **temperature**. **Nonmetals** are very poor conductors and are used for thermal **insulation**.

conservation of energy The principle that the total **energy** in any system is constant. It is a basic law of physics that energy cannot be created nor destroyed. In any energy transfer, the output energy is equal to the input energy (see **efficiency**). This law was shown to be incomplete by Einstein who proved that an object's **mass** and energy are related. The relationship may be expressed by the following equation:

$$\text{Energy change} = \text{mass change} \times \text{speed of light}^2$$
$$E = m \times c^2$$

The energy released in both nuclear **fission** and nuclear **fusion** is the result of mass changes in atomic nuclei. The *law of conservation of mass and energy* is that the sum of an object's mass and energy is constant unless there is transfer to or from outside.

convection The transfer of **energy** through a fluid by movement within the **fluid**. An increase in the **temperature** of a fluid increases its **volume**. In turn, an increase in volume causes a decrease in **density** (since density = **mass**/volume). Thus when a fluid gets warm it rises and displaces colder, more dense fluid.

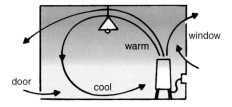

convection *How a convector heater heats a room.*

This sets up a convection current which circulates **heat**.

Winds and breezes are convection currents on a huge scale in the **atmosphere**.

cooling curve The general name given to a graph of **temperature** taken as a substance cools down over a period of time. The shape of the graph indicates when a change of state has taken place (see **latent heat**). It is particularly useful for finding the **melting point** of a **liquid** – the same temperature at which the liquid becomes solid. It may also be used to find the heat or temperature losses due to cooling.

copper A transition **metal** which plays an important part in our lives. It is a vital trace element in our bodies and we need between 1–2 mg per day.

Copper comes low down in the **electrochemical series** and is an unreactive metal. It is sometimes found as the free (native) metal which is an indication of its lack of reactivity. It does not produce **hydrogen** when added to dilute acids.

Copper is obtained from its sulphide ores chalcopyrite ($CuFeS_2$) and bornite (Cu_5FeS_4), and its carbonate ores malachite ($CuCO_3$. $Cu(OH)_2$) and azurite ($2CuCO_3$. $Cu(OH)_2$). The final stage of purification is achieved by **electrolysis**. Impure metal is used for the *anode*, pure copper for the *cathode* and copper(II) sulphate as the **electrolyte**. Copper is widely used in plumbing and for electrical wiring.

corrosion The eating away of a substance by chemical reactions with other substances. Good examples are:
(a) The **weathering** of limestone buildings by rainwater which contains **acids**.
(b) The rusting of **iron** and **steel** due to **oxidation** in the presence of **air** and moisture.

Corrosion begins on the surface and often a surface layer is formed which protects the rest of the material. In rusting this is not the case. The rusting process goes through iron and some steels until it is all corroded. *See* **rust**.

cost of electricity The cost of operating an electrical device is determined by its power, in kilowatts (kW), the length of time it is used, in hours (h) and the unit price of electricity, in kilowatt hours (kWh):

$$\text{Cost} = \text{power (kW)} \times \text{time (h)} \times \text{price of 1kWh}$$

See **electricity meter**.

covalent bonds This type of **bond** is formed when two atoms come together and share **electrons**, e.g. **hydrogen atoms** have one electron each. The atoms in a **molecule** of hydrogen share both electrons between them.

hydrogen atoms (H)   hydrogen molecule (H_2)

covalent bonds *A molecule of hydrogen.*

Each shared pair of electrons produces a covalent bond. Covalent bonds are usually found in **compounds** which only contain nonmetallic **elements**, e.g. **carbon dioxide** (CO_2), **ammonia** (NH_3), hydrogen chloride (HCl), and elements which exist as molecules which contain two or more atoms, e.g. **hydrogen** (H_2), **oxygen** (O_2), nitrogen (N_2), **chlorine** (Cl_2). Covalent bonds which contain two electrons, i.e. one shared pair, are called single bonds. Many molecules contain *double* or *triple* bonds.

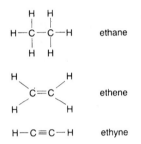

covalent bonds   *Molecular structures containing single, double and triple bonds.*

covalent compounds Compounds which contain **covalent bonds**. They tend to be non conductors and have low **melting** and **boiling points**.

current (I) The flow of **charge** through an electrical conductor (see **conduction, electrical**). The unit of current is the ampere, A; often abbreviated to amp. One amp is the current when a charge, Q, of one coulomb passes in one second, $I = Q/t$.

Currents in **circuits** are the basis of many electrical devices and electronic systems. In each case the effects of charge and **energy** transfer are used. Energy transfer may produce **light**, **temperature** change, magnetic effects, **chemical changes** and mechanical **force**.

D

decay In physics this word is used to describe how the size of a measure decreases with time. In **radioactivity** it describes how the activity of a substance decreases with time. *See* **half-life period**.

decomposition The breaking up of a **compound** into other compounds and **elements**. **Heat** is sometimes used to bring this about (thermal decomposition). Some examples are:

$$2Cu(NO_3)_2(s) \rightarrow 2CuO(s) + 4NO_2(g) + O_2(g)$$

copper(II)   copper(II)   nitrogen(IV)   oxygen
nitrate   oxide   oxide

$$2HgO(s) \rightarrow 2Hg(l) + O_2(g)$$

mercury   mercury   oxygen
oxide

$$CaCO_3(s) \rightarrow CaO(s) + CO_2(g)$$

calcium   calcium   carbon
carbonate   oxide   dioxide

denitrification This process occurs in **soil** and involves the conversion of nitrates into nitrogen by denitrifying bacteria. Denitrification causes a decrease in the fertility of the soil as the nitrogen is lost to the **atmosphere**. *See* **nitrogen cycle**.

density The density of a substance is the **mass** of a sample divided by its **volume**. The **SI unit** is the kilogram per cubic metre kg/m^3, although density is also often given in g/cm^3:

$$\text{density } (kg/m^3) = \frac{\text{mass (kg)}}{\text{volume } (m^3)}$$

The density of water is $1000 \ kg/m^3$ (this is equal to $1 \ g/cm^3$).
As **solid** and **liquid** volumes change little with changes of **temperature** and **pressure** their density is fairly constant. By contrast, the density of a **gas** varies widely with temperature and pressure.

dental formula A formula which describes the number and type of **teeth** in the jaws of an adult mammal with a complete set of teeth. It is expressed by writing the number of teeth in the upper jaw on one side of the mouth over those in the lower jaw on the same side.
It is not necessary to give a formula for the whole jaw as the teeth of mammals are always symmetrical about the centre of the jaw. In humans

the upper and lower jaws are also symmetrical, however this is often not the case for other mammals.

deposition The process by which agents of **erosion** drop the fragments of rock which they carry along. *See* **rock cycle**.

detergent A cleaning agent used in washing articles such as clothes and dishes. During washing a detergent acts in two ways:
(a) It reduces the surface tension of the water thus allowing the water to wet things more thoroughly.
(b) It brings together the water and (normally insoluble) fat, oil or grease to form an emulsion.

If the presence of a detergent is accompanied by rapid movement, such as the washing action of a washing machine, then particles of dirt and grime are removed and the article is cleaned.

Detergents are able to form an emulsion because one part of the **molecule** is ionic and another part (the **hydrocarbon** chain) is covalent. The ionic part is attracted to water while the covalent part is attracted to oil. Thus detergent molecules hold the water and oil together.

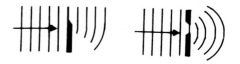

detergent *A synthetic (soapless) detergent molecule.*

Soaps work in a similar way to detergents; however their action is generally less vigorous. Soaps are made from natural oils and are sodium salts of long chain carboxylic acids. Detergents are made from **petroleum**.

dialysis A process in which small **molecules** are separated from larger ones using a *semipermeable membrane*. Only the smaller molecules are able to pass through the membrane in an excess of water. It is by this process that the **kidneys** are able to purify the **blood**. The principal is also the basis for kidney machines, used to purify the blood in cases of kidney disease or *failure*.

diffraction The bending of a wave round the edge of an opaque object into the region of shadow. The effect of diffraction on water waves can be seen in harbours or with a ripple tank.

All waves can diffract around

diffraction *The diffraction of water waves seen in a ripple tank.*

suitable objects. The effect is greatest when the size of the object is about the same as the wavelength of the wave.

Light diffracted by a narrow slit produces a set of interference fringes. As diffraction depends on wavelength, it can produce **dispersion**. A diffraction grating can be used to produce spectra.

diffusion This is the way in which **fluid** particles spread from a source through the space available. For example, if a **gas** with a distinct odour (such as hydrogen sulphide) is released in the corner of a room it takes very little time before people all over the room will be able to smell it. Diffusion in **liquids** is not as fast as in gases, because the particles move at slower speeds and collide more often. Diffusion is part of the evidence for the **kinetic model** of matter.

digestion The breakdown of large insoluble food particles into small soluble particles by the action of **enzymes**. This occurs prior to **absorption** (*see* **absorption 2**) and **assimilation**. In many animals, including mammals, digestion and absorption take place in the **alimentary canal**. The table on page 61 gives some details of the digestive enzymes found in the human alimentary canal.

Location	Glands	Enzyme	Substrate	Product
Mouth	Salivary	Amylase	Starch	Maltose
Stomach	Gastric	Pepsin	Protein	Peptides
		Rennin	Milk protein	Coagulated milk
Duodenum	Pancreas	Amylase	Starch	Maltose
		Lipase	Fats	Fatty acids + glycerol
		Trypsin	Protein + peptides	Amino acids
Ileum		Lactase	Lactose	Glucose + galactose
		Lipase	Fats	Fatty acids + glycerol
		Maltase	Maltose	Glucose + fructose
		Peptidase	Peptides	Amino acids

digestion *Digestive enzymes in humans.*

diode A two-electrode device which will allow an electric current to pass in one direction only.

The main use of diodes is as **rectifiers**. Older types of diode include diode valves and the point-contact diode ('cat's whisker').

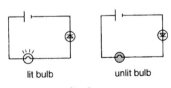

lit bulb   unlit bulb

diode

The modern p–n junction semiconductor diode has now replaced other types for almost all applications.

direct current (DC) A flow of **charge** in one direction only. Direct current may be constant or may vary in value.

dispersion The separating of **radiation** into bands of different wavelengths by bending it (**refraction** or **defraction**). The angle of bending depends on the wavelength. For example, *white light* consists of light rays of different wavelengths (*see* **colour**). When white light passes from **air** into glass, the rays bend progressively, going from the reds to the blues. The white light is dispersed.

dissociation The breaking up of a **compound** into smaller **molecules** or **ions**. This dissociation is reversible:

$$HCl(aq) \quad \xrightarrow[\text{heat}]{} \quad H^+(aq) \quad + Cl^-(aq)$$

$$NH_4Cl(s \quad \rightarrow \quad NH_3(g) \quad + HCl(g)$$

distance-time graph This shows how the distance an object travels varies with time. If the object moves at a constant **speed** the graph will be a straight line. A curve indicates that the speed of the object is not constant, but varies.

The speed of the object is given by the gradient or slope of the graph, y/x. Where speed is constant the gradient of the graph is the same at any point. However, where speed varies the gradient of the graph changes and must be worked out for each point.

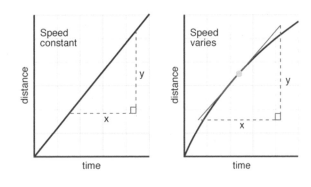

dm³ The symbol for the cubic decimetre. The cubic decimetre is used by scientists to measure **volume**. In everyday life, volume is measured using another unit, the litre. One litre is exactly equal to one cubic decimetre:

$$1000 \text{ cm}^3 = 1 \text{ dm}^3 = 1 \text{ litre}$$

DNA (deoxyribonucleic acid)

A major constituent of **genes** and hence **chromosomes**.

A *polynucleotide* chain consists of a series of alternating **sugar** and phosphate groups with *nitrogen bases* attached to the sugar groups.

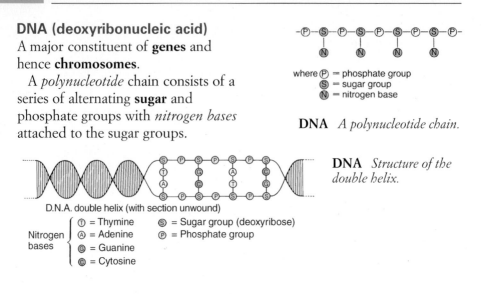

where Ⓟ = phosphate group
Ⓢ = sugar group
Ⓝ = nitrogen base

DNA   *A polynucleotide chain.*

DNA   *Structure of the double helix.*

D.N.A. double helix (with section unwound)

Nitrogen bases
Ⓣ = Thymine   Ⓢ = Sugar group (deoxyribose)
Ⓐ = Adenine   Ⓟ = Phosphate group
Ⓖ = Guanine
Ⓒ = Cytosine

DNA consists of a double polynucleotide chain twisted into a helix. The two chains are held together by bonds between nitrogen base pairs. These nitrogen bases can only link as complimentary pairs: thymine with adenine, and guanine with cytosine. The number and sequence of the base pairs in the DNA polynucleic chain represents coded information (the *genetic code*) which allows the transfer of hereditary information from generation to generation.

drug   A substance taken for the effect that it has on the body or for the prevention or treatment of disease. Main groups of drugs include (a) stimulants, such as caffeine, that increase body activity, (b) sedatives, such as barbiturates, that have a claming effect on the body, and (c) analgesics or pain killers, such as paracetamol, taken to reduce pain.

Drug abuse results from the habitual consumption of substances that effect the nervous system. Such drugs include heroin, cocaine, ecstasy LSD and solvents (glue and solvent sniffing). The body may become physiologically dependent on these drugs leading to *drug addiction* and, in some cases, eventual death. Overdosing on any drug may result in death.

E

ear The **organ** responsible for hearing and balance in vertebrates. Hearing is a sensation which is produced by vibrations or **sound waves**. These are converted into nerve impulses by the ear and transmitted to the **brain**, where they are interpreted.

The ear is often considered in terms of three sections:

(a) *Outer ear.* The pinna is the part of the ear which we can see. It is funnel-shaped and directs sound waves into the ear and along the auditory canal. At the end of the canal there is a thin membrane, the eardrum (tympanum), which is made to vibrate by the sound waves.

(b) *Middle ear.* This is an air-filled cavity. It is connected to the back of the mouth (pharynx) by the eustachian tube. This allows **air** into the middle ear so that the air **pressure** on each side of the eardrum is always the same. Within the middle ear there are three tiny **bones** called ossicles. The

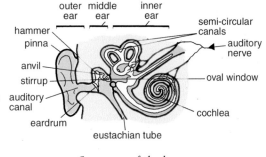

ear *Structure of the human ear.*

individual bones are named from their shape; malleus (hammer), incus (anvil) and stapes (stirrup). The ossicles transmit and amplify the vibrations of the eardrum to a membrane called the oval window which lies between the middle and the inner ear.

(c) *Inner ear.* This is filled with **fluid** and contains the cochlea and semi-circular canals. The vibration of the stapes against the oval window sets up waves in the fluid of the cochlea. These waves stimulate receptor cells (hair cells) which result in impulses being sent to he brain via the auditory nerve. Within the brain these impulses are interpreted as sounds.

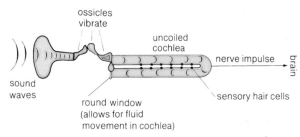

ear *Sound waves are converted into nerve impulses.*

Balance is maintained by the semicircular canals, in association with information received from the **eyes** and **muscles**. The canals contain fluid and receptor **cells**. The cells are stimulated by movement of the fluid during changes of posture. The nerve impulses initiated by these cells travel to the brain along the auditory nerve and trigger the response needed for the body to retain normal posture.

earth 1. To join an object to earth by a conductor so that it will share any net **charge** it has with the earth. Because the earth is so large, this effectively means that the earthed object cannot keep charge. To earth an object it is connected to a metal plate or stake in the ground.

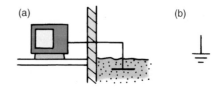

earth *(a) An object is earthed by connecting it to a metal plate in the ground; (b) the standard symbol for an earth lead.*

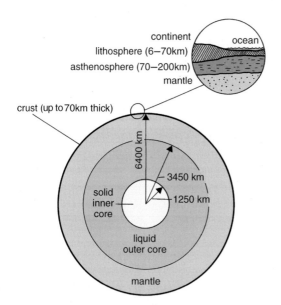

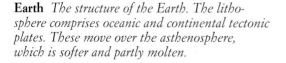

Earth *The structure of the Earth. The lithosphere comprises oceanic and continental tectonic plates. These move over the asthenosphere, which is softer and partly molten.*

The green/yellow coated wire in a household cable should be connected to the earth pin of the **three-pin plug**. In the event of a fault in an appliance the earth wire protects the user. Any charge flows to earth through the wire, and not through the person who touches the appliance.

2. The Earth is the **planet** upon which we live. It consists of a central core surrounded by a very dense liquid mantle. Surrounding the mantle, and floating on it, is the earth's crust.

earthquake A shaking due to **energy** released by a sudden fracturing of the earth. The energy is released in the form of three types of **wave**:
(a) P-waves – push and pull, or primary waves.
(b) S-waves – shake, or secondary waves.
(c) L-waves – long waves.
 The place within the earth where an earthquake starts is called the *focus*. The point on the earth's surface directly above the focus is called the *epicentre*. The magnitude of an earthquake is the amount of energy which is released. This is measured on the *Richter scale*. Earthquakes often occur at places near plate boundaries (*see* **tectonic theory**). Active **volcanoes** are often located in earthquake zones.

echo An effect caused by **reflection**. **Radiation** from a source appears to come, after a short delay, from somewhere else, the image. The word is most often used in connection with sound.

eclipse To prevent **light** from a source reaching an object. Light travels in a straight line, through a given medium, from a source to an object. A second *opaque* object, placed in such a way as to prevent light from reaching the first object, is said to eclipse, or hide, it.

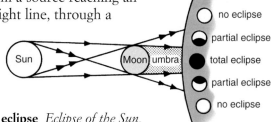

eclipse *Eclipse of the Sun.*

The two most common types are:
(a) Eclipse of the **Sun** (solar eclipse). If the **Moon** passes exactly between the **Earth** and the sun its **shadow** passes over part of the Earth's surface. People in the shadow see a solar eclipse. The eclipse is total for people in the umbra (full shadow) and partial for people in the penumbra (partial shadow).
(b) Eclipse of the Moon (lunar eclipse). When the Earth is situated in a line between the Sun and the Moon a lunar eclipse occurs. The Moon is invisible from Earth when passing through the Earth's umbra and partially visible when passing through its penumbra.

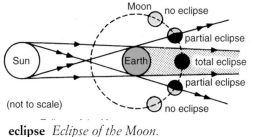

eclipse *Eclipse of the Moon.*

ecosystem A natural unit consisting of living parts, in the form of a **community** of plants and animals; and non-living parts, in the form of a **habitat**:

$$\text{habitat} + \text{community} \rightarrow \text{ecosystem}$$

Ecosystems may be forests, grasslands, lakes, rivers, etc. The driving force of all ecosystems is the energy provided by the **Sun**.

efficiency The measurement of the ratio of the useful **energy** output to the total energy input in any energy transfer. It is often given as a percentage and has no units:

$$\text{efficiency} = \frac{\text{work (energy) output}\ (\times 100)}{\text{work (energy) input}}$$

Of course, none of the input energy is destroyed; some is simply transferred into unwanted forms, e.g. a car engine transfers chemical energy into mechanical energy with about 25% efficiency. The majority of the energy wasted is in the form of unwanted heat.

elasticity The property of a substance to resist being deformed when a force is applied to it and to resume its original shape when the force is removed. However, if the applied force is greater than the *elastic limit* the substance will be permanently deformed and becomes a **plastic**. These effects are the result of stress/strain between the particles.

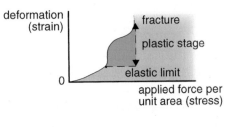

elasticity *The stress/strain curve of a metal sample.*

All elastic substances obey **Hooke's law**.

electricity current electricity is a moving electric **charge** in a **circuit**. As charge moves around a circuit **energy** is transferred. This form of electricity is a convenient way of transferring energy and is widely used to power a variety of devices. (*See also* **static electricity**).

electricity meter A device which measures the amount of electrical **energy** used by a consumer. The unit used is the *kilowatt-hour*, kWh. (1 kWh is equal to 3.6×10^6 J). Older types of meter show the reading on a series of dials, corresponding to different powers of 10. Modern types have a single digital reading.

Domestic *electricity bills* are normally sent out to consumers every three months. They consist of a rental, for the electricity meter, and a charge for the number of units of electricity used. *See* **cost of electricity**.

electrochemical series A series in which the **elements** are arranged in order of their *standard oxidation potentials*. The following list gives some common elements in order of decreasing oxidation potential:

Potassium	2.92
Calcium	2.76
Sodium	2.71
Magnesium	2.38
Zinc	0.76
Hydrogen	0.00
Copper	−0.34
Iodine	−0.54
Bromine	−1.06
Gold	−0.80
Chlorine	−1.36
Silver	−1.42
Fluorine	−2.87

This series is useful because it allows us to compare reactivities and make predictions about whether certain reactions will or won't occur. An element will only displace less reactive elements from their **compounds**. Here are some examples:
(a) Zinc will remove the oxygen from copper oxide but copper will not remove the oxygen from zinc oxide.
(b) Hydrogen will reduce copper(II) oxide but not zinc oxide.
(c) Copper will not react with **acids** to release hydrogen. *See* **reactivity series**.

electrode A conductor (*see* **conduction 1.**) which dips into an **electrolyte** and allows **current** to flow to and from the electrolyte. **Copper, carbon** and platinum are commonly used as electrodes in **electrolysis**.
During electrolysis, the electrode connected to the positive pole of the electrical supply is called the anode and the electrode connected to the negative pole is called the cathode. Chemical reactions occur at both electrodes.

electrolysis The result obtained when a direct (electric) **current** is passed through a liquid which contains **ions** (an **electrolyte**) and **chemical changes**

occur at the two **electrodes**.
For example, the electrolysis
of sodium chloride solution:
This electrolysis of sodium
chloride is carried out on a large
scale industrially to make
sodium hydroxide. Electrolysis
is used to extract **metals** and
nonmetals from their
**compounds. Aluminium,
sodium**, and **copper** are
produced in this way, as is
chlorine.

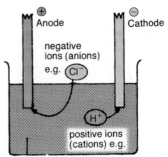

Electrolyte (sodium chloride solution)

electrolysis *The electrolysis of sodium chloride solution.*

At the anode	At the cathode
$Cl^- \rightarrow Cl + e^-$	$H^+ + e^- \rightarrow H$
$2Cl \rightarrow Cl_2$	$2H \rightarrow H_2$
chlorine is given off	hydrogen is given off

electrolysis *The electrolysis of hydrochloric acid.*

electrolyte An electrolyte is either:
(a) A molten **ionic compound**:

$$NaCl(l), \ PbBr_2(l)$$

(b) A solution which contains **ions**:

$$HCl(aq), \ NaOH(aq), \ CuSO_4(aq), \ NaCl(aq)$$

Chemical changes take place at the electrodes when a direct electric
current is passed through an electrolyte. This process is called **electrolysis**.
With molten compounds, such as sodium chloride, the changes are simple:

$$2Na^+Cl^-(l) \rightarrow 2Na(l) + Cl_2(g)$$

With ionic solutions the changes are more complicated. The products
depend on:
(a) The **concentration** of the solution.
(b) The **voltage** which is applied.
(c) The type of **electrode** which is used.
(d) The nature of the ions present in the solution.
The following example shows what happens when copper sulphate
solution is electrolysed using different types of anode:

	Anode	Reaction
(i)	**carbon** or platinum	$4OH^-\,(aq) \rightarrow 2H_2O(l) + O_2(g) + 4e^-$
(ii)	**copper**	$Cu(s) \rightarrow Cu^{2+}(aq) + 2e^-$

In process (i) oxygen is released, while in process (ii) the copper anode dissolves. Process (ii) is used in the purification of copper.

electromagnet

A device which acts as a **magnet** only when its coil is carrying an electric **current**. In the school laboratory simple electromagnets are made by winding several coils of wire around an **iron** nail and connecting the ends of the wire to a **battery**. Much larger versions, based on the same principle, are used in places like scrap yards for separating iron and **steel** from other materials.

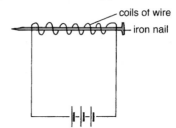

electromagnet   *A simple electromagnet.*

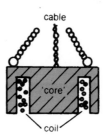

electromagnet   *An industrial electromagnet.*

Electromagnets are also used in a variety of electrical devices such as **bells**, **relays** and speakers.

electromagnetic induction

The generation of a **voltage** across a conductor when it moves with respect to a **magnetic field**. This induced *electromotive force* (*emf*) is increased if there is an increase in:

(a) The relative motion between the conductor (usually a coil) and the magnetic field.

(b) The length of the conductor (number of coils of wire) in the magnetic field.

(c) The strength of the magnetic field.

The direction of the induced voltage is such as to oppose the change which is producing it: this is known as *Lenz's law*. Electromagnetic induction is the basis of many electrical devices such as the **generator** and the **transformer**.

electromagnetic waves A range of **radiations** which differ in wavelength but have the following properties in common:
(a) They are produced by moving electric **charge**.
(b) They travel by vibrating electric and magnetic (hence electromagnetic) fields.
(c) They move through empty space at the speed of **light** (300 000 000 m/s).
(d) They tend to be absorbed by matter in which they travel more slowly.
(e) Like all waves they show **reflection, refraction,** *interference* and *diffraction* effects.
(f) They are *transverse waves* and show polarization effects.
(g) They can show particle properties.

   The **spectrum** of electromagnetic waves is usually divided into a number of regions. The following table shows approximate values for the wavelengths and **frequencies** of the different regions:

Region	Wavelength (m)	Frequency (Hz)
gamma	-10^{-12}	$10^{21}-$
X-ray	$10^{-12}-10^{-10}$	$10^{18}-10^{21}$
ultraviolet	$10^{-10}-10^{-7}$	$10^{15}-10^{18}$
visible light	$10^{-7}-10^{-6}$	$\approx 10^{15}$
infrared	$10^{-6}-10^{-3}$	$10^{12}-10^{15}$
microwave	$10^{-3}-10$	$10^{7}-10^{12}$
radio	$10-10^{6}$	$10^{2}-10^{7}$

electromagnetic waves *Electromagnetic wave regions.*
Note – all figures are approximate and some regions overlap.

electron A very small subatomic particle. Electrons carry a negative **charge** and move around the nuclei of an **atom. Ions** are formed when atoms lose or gain electrons, i.e:

   gaining electrons – negative anions, e.g. Br^-, Cl^-, O^{2-}, S^{2-}

   losing electrons – positive cations, e.g. Al^{3+}, Fe^{2+}, H^+, Na^+

   The way in which electrons are located in an atom is called the electronic configuration. An electric **current** is a flow of electrons moving through a conductor.

element A pure substance which cannot be broken down into anything simpler by chemical means. There are 92 elements which occur naturally on the earth, and a small number of others have been made artificially in laboratories by nuclear reactions. All elements have a unique number of **protons** in their **atoms**.

embryo **1.** A young animal developed from a zygote as a result of repeated **cell** division. During **pregnancy** in mammals, the embryo develops within the female **uterus**. When the main features of the embryo have developed it is called a **fetus**.

  2. A young flowering plant developed from a fertilized ovule. In seed plants this is enclosed within a **seed** prior to **germination**.

empirical formula The formula of a **compound** which shows the different atoms that are present in the molecule in their simplest whole number ratio. Here are some examples:

Compound	Molecular formula	Empirical formula
Ethene	C_2H_4	CH_2
Butane	C_4H_{10}	C_2H_5
Ethanoic acid	$C_2H_4O_2$	CH_2O

  In many compounds the empirical and molecular formulae are the same. *See* **molecular formula**.

endocrine glands (or **ductless glands**). Structures found in vertebrates and some invertebrates. They release chemicals called **hormones** directly into the bloodstream. The rate at which the hormones are secreted is often a response to changes in internal body conditions; however, it may also be a response to environmental changes. The major endocrine glands in the human body are the pituitary gland (at the base of the brain), the thyroid gland (in front of the trachea), the **pancreas**, the **ovaries** (in females), the testes (in males), and the adrenal glands (above the kidneys).

endothermic reaction A reaction in which **heat energy** is taken in from the surroundings to make the reaction occur; e.g. the thermal decomposition of calcium carbonate:

$$CaCO_3(s) \rightarrow CaO(s) + CO_2(g)$$

  In endothermic reactions there is more energy in the **bonds** of the products than there was in the bonds of the starting materials. Compare **exothermic reaction**.

energy (W) The ability of an object to do *work*. The unit of energy is the **joule**, J. It is often useful to speak of different 'forms of energy' (although what is really meant is energy in different contexts):

(a) *Gravitational potential* energy: the work a **mass**, m, can do by falling a distance, h. $W=mgh$.

(b) *Electrical energy*: the work an electric **current**, I, can do in time, t. $W= VIt$.

(c) **Kinetic energy**; the work a mass, m, moving at speed, v, can do in coming to rest. $W= (1/2)\ mv^2$.

(d) *Thermal energy*: the work needed to raise a mass, m, of *specific heat capacity*, c, by θ degrees. $W= mc\theta$.

In all energy transfers the useful energy output is less than the total energy input because some of the input energy will be transferred into unwanted

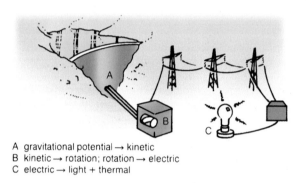

A gravitational potential → kinetic
B kinetic → rotation; rotation → electric
C electric → light + thermal

energy *A chain of energy transfers.*

forms. The ability of a device to transfer energy from one form to another is measured by its **efficiency**.

engines Systems in which the chemical (potential) **energy** of a **fuel** is transferred to mechanical energy (work). The **combustion** of the fuel in **oxygen** results in a large volume of hot **gas**. As the **volume** of gas increases it transfers energy to the system.

environment A collective name for the conditions in which organisms live. Many factors contribute to the environment including:

(a) Nonliving physical (abiotic) factors such as **temperature** and **light**.

(b) Living (biotic) factors such as *predators* and *competition*.

The interaction of all these factors determines the conditions within **habitats** and selects the **communities** of organisms which are best suited to the prevailing conditions.

enzymes **Protein** substances which act as **catalysts** within **cells**. Catalysts are substances which control the rates of chemical **reactions**. In a cell there may be hundreds of chemical reactions occurring at the same time. Each of them will require its own particular enzyme as enzymes are normally very specific in their action and will only catalyse one particular reaction.

equation   45

Enzymes may catalyse either:

(a) *Synthesis reactions* in which complex **compounds** are produced from simple **molecules**.

(b) *Degradation reactions* in which complex molecules are broken down into simple subunits by hydrolysis.

Enzymes work most efficiently within a narrow **temperature** range. Thus human enzymes work best around human body temperature (37 °C). Efficiency decreases above and below this temperature and at temperatures above 45 °C most enzymes are destroyed.

Enzymes work best at a particular **pH**. This varies for different enzymes:

(a) *Salivary amylase*, found in the saliva, works best at a neutral or slightly acid pH.

(b) *Pepsin*, found in the stomach, will only work in an acid pH.

(c) *Trypsin*, found in the intestine, works best in an alkaline pH.

The rate of an enzyme-catalysed reaction increases as the enzyme *concentration* increases.

The rate of an enzyme-catalysed reaction also increases as the concentration of the substance on which the enzyme acts increases up to a maximum point.

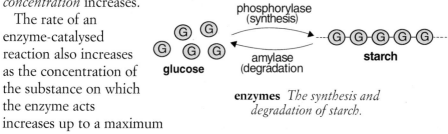

enzymes *The synthesis and degradation of starch.*

Most enzymes are named by adding the suffix *-ase* to the name of the substance the enzyme acts on. For example maltase is the enzyme which acts on maltose.

equation A means of expressing a chemical reaction. An equation may be in words:

$$\text{hydrogen} + \text{oxygen} \rightarrow \text{water}$$

or it may be a formula (or symbol):

$$H_2 + O_2 \rightarrow H_2O$$

An equation is balanced if there are the same number of each type of **atom** on each side of the equation, e.g:

$$2H_2 + O_2 \rightarrow 2H_2O$$

This is a balanced equation, because there are four **hydrogen** atoms and two **oxygen** atoms on each side of it. The equation tells us that two molecules of hydrogen react with one molecule of oxygen to produce two molecules of **water**. Using the equation and the relative atomic masses of hydrogen (1) and oxygen (16), we can further state that 2 **moles** or 4 g of hydrogen reacts with 1 mole or 32 g of oxygen to produce 2 moles or 36 g of water.

$$2H_2 + O_2 \rightarrow 2H_2O$$
$$2(2) + 32 = 2(2 + 16) = 36$$

Sometimes *ionic equations* are used:

$$Cu^{2+} + Zn \rightarrow Cu + Zn^{2+}$$

State symbols are often added after the formula to denote which **state of matter** a substance is in, e.g. hydrogen and oxygen **gases** react to produce **liquid** water.

$$2H_2(g) + O_2(g) \rightarrow 2H_2O(l)$$

equilibrium A state of balance between opposing reactions. In a **reversible reaction** the reaction proceeds in both directions, e.g.

$$3Fe(s) + 4H_2O(g) \rightleftharpoons Fe_3O_4(s) + 4H_2(g)$$

The reactants, iron and steam, produce the products, iron oxide and **hydrogen**. As soon as the products are made they begin to react together to reform iron and steam. Eventually, a situation is reached where the rate of the forward $\rightarrow$ reaction equals the rate of the reverse $\leftarrow$ reaction. At this point the proportion of the four substances present is constant and there appears to be no reaction taking place. The forward and reverse reactions are in dynamic equilibrium.

erosion The removal of rock and mineral fragments by **water**, wind and ice. The processes of **weathering** and erosion operate together. Their combined effect is known as *denudation*.

erythrocyte *See* **red blood cell**.

ethanol One of a group of *alcohols*. It is a colourless flammable **liquid** with a **boiling point** of 78 °C. Ethanol is the alcohol in alcoholic drinks. In humans, small amounts relax the body but large amounts lead to alcohol poisoning and even death.

C_2H_5OH

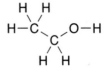

ethanol

Alcoholic drinks are made by **fermentation**. Some of the ethanol produced by industry is also made this way, however most is made by the hydration of ethene:

$$C_2H_4(g) + H_2O(l) \rightarrow C_2H5_2OH(l)$$
ethene   water   ethanol

Ethanol is used in industry as a solvent and is the main constituent of methylated spirits.

evaporation The change of state from **liquid** to **gas** which can occur at any **temperature** up to the **boiling point**. At any one time, a variable population of **molecules** in a liquid will have sufficient **energy** to escape into the **atmosphere**. If a liquid is left in an open container for long enough it will all evaporate. In general, the lower the boiling point, the faster the rate of evaporation, though this also depends on the **latent heat**: Liquids which are easily turned into gases are said to be volatile.

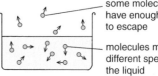

some molecules have enough energy to escape

molecules moving at different speeds in the liquid

evaporation *Molecules escaping from the body of a liquid into the atmosphere.*

evolution The development of complex organisms from simpler ancestors over successive generations. *See* **natural selection**.

excretion The process by which organisms get rid of the waste products of **metabolism**. The main excretory products are **water**, **carbon dioxide** and nitrogenous compounds such as **urea**.

In simple organisms excretion occurs through the **cell** membrane or epidermis. In higher plants excretion occurs through the leaves. Most animals have specialized excretory organs. For example, in humans the **lungs** excrete water and carbon dioxide and the **kidneys** excrete urea.

exothermic reaction A reaction in which **heat energy** is released from the reactants to the surroundings. The bonds of the products contain less energy than the bonds of the reactants, hence the products are more stable than the reactants. Many common reactions are exothermic. All **combustion** and **neutralization** reactions are exothermic. Some important industrial processes are also exothermic, e.g:

Haber process: $3H_2(g) + N_2(g) \rightarrow 2NH_2(g)$ *See* **ammonia**.

Contact process: $2SO_2(g) + O_2(g) \rightarrow 2SO_3(g)$ *See* **sulphuric acid**.

Compare **endothermic reaction**.

expansion The change in size of an object or material due to **temperature** change. In terms of the **kinetic model** a temperature rise causes a rise in the particles mean **energy**: i.e. the particles move more quickly and try to take up more space. This effect is greatest with **gases**. All gases expand by roughly the same amount for a given temperature change. **Solids** and **liquids** expand much less but vary greatly in the amount of expansion. *See* **bimetallic strip, water**.

Thermal expansion has many uses such as in **thermometers** and **thermostats**. It can also cause problems in structures such as clocks, roads and bridges.

eye A *sense organ* which responds to **light.** Eyes vary in complexity from the simple structures found in invertebrates to the complex eyes of insects (*compound eye*) and vertebrates.
(a) The sclerotic layer is a tough protective layer which surrounds the outside of the eye. At the front of the eye it forms the transparent cornea.
(b) The choroid layer is beneath the sclerotic layer. It contains black pigmentation which prevents reflection within the eye. It is rich in blood vessels supplying the eye with food and oxygen.

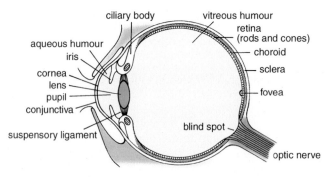

eye *Vertical section through the human eye.*

(c) The retina is the inner layer of the eye. It is a layer of nerve **cells** which are sensitive to light. There are two types of cell named from their shape.
  (i) *Rods* are very sensitive to low light intensity. The eyes of nocturnal animals have high concentrations of this type of cell.
  (ii) *Cones* are sensitive to bright light and some are stimulated by light of different wavelengths hence they are responsible for colour

vision. The *fovea* is a small area of the retina at the back of the eye which has many cones but no rods. It gives the greatest degree of detail and colour.

(d) The *blind spot* is the part of the retina where the nerve fibres, which are connected with the rods and cones, leave the eye and enter the *optic nerve*. There are no light-sensitive cells at the blind spot, hence an image here is not registered by the brain.

(e) The *aqueous humour* and *vitreous humour* are fluids contained in the eye. They help to maintain the shape of the eye, play a small part in focusing the light and allow

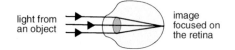
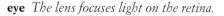

light from an object → image focused on the retina

eye *The lens focuses light on the retina.*

nutrients, oxygen and wastes to diffuse (*see* **diffusion**) into and out of the eye cells.

(f) The *lens* is a transparent *biconvex* structure which can change curvature. It is mainly responsible for focusing light on the cells of the retina.

Accommodation is the ability of the eye to focus on objects of varying distances from it. This is possible by altering the curvature of the lens.

The lens is held in place by *suspensory ligaments* which are attached to *ciliary muscles*. When the ciliary muscles are relaxed the **pressure** of the fluid in the eye keeps the suspensory ligaments taut so the lens is pulled out and its centre is thin. This is the normal relaxed state of the eye when looking at distant objects. To look at closer objects, the ciliary muscles contract. The suspensory ligaments become slack and the lens goes fatter at its centre.

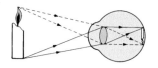

eye *The image is smaller than the object, and is inverted.*

The amount of light entering the eye is controlled by the *iris*. This is the coloured part of the eye and contains muscles. The hole in the centre of the iris through which light enters is called the *pupil*. In poor light the pupils are wide open (dilated) to allow the maximum amount of light into the eye. As the light gets brighter the pupil becomes smaller (contracts). This protects the retina from possible damage. This mechanism is an example of a **reflex action**. Because the pupil is small the light rays enter the eye in such a way that they produce an image which is upside down (inverted) on the retina. This is corrected by the brain.

Long sight and short sight are two common eye defects which can be corrected by **lenses**:

(a) A person with long sight cannot see close objects clearly because the light focuses behind the retina. This can be corrected by wearing converging (convex) lenses.

(b) A person with short sight cannot see distant objects clearly because the light focuses in front of the retina. This can be corrected by wearing diverging (concave) lenses.

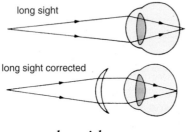

long sight

F

fats or lipids *Organic compounds* which contain the **elements carbon, hydrogen** and **oxygen**. Fats are made up of three *fatty acid* **molecules** (which may be the same or different) bonded to one glycerol molecule. Fat deposits under the skin act as a long-term **energy** store. 1 g of fat contains 39 kJ of energy. These deposits also provide **heat insulation**.

Fat is also an important constituent in the **cell** membrane. Its insolubility in **water** is utilized in the waterproofing systems of many organisms.

feedback A method of control in which part of the output signal is returned as the input signal.

An *amplifier* is a device in which the output signal has the same form as the input, but is greater in amplitude or size. The output/input amplitude ratio is called the *gain*. In an amplifier, feedback may be used to control the gain.

In an industrial process feedback is used to control the quality of the product. In the following example the thickness of paper is monitored by measuring the quantity of **beta particles** which passes through it.

If the number increases the paper is too thin, if it decreases the paper is too thick. This information is continually fed back to the controller which operates the rollers. This adjusts the pressure automatically in order to keep the paper at a preset thickness.

Many biochemical processes are controlled by feedback; the product of a process controls that process. For example, the thyrotrophic hormone (TSH), secreted by the pituitary gland of the **brain**, controls the amount of the **hormone** thyroxine which is produced by the thyroid gland. A high concentration of thyroxine in the body suppresses the release of TSH, thus less thyroxine is made. When the thyroxine level in the body is low, more TSH is released causing more thyroxine to be pro duced. Other examples of feedback in the body include control of the body water level by the antidiuretic hormone (ADH) and control of the blood sugar level by the hormone insulin. (*See* **kidney**.)

control *A simple control system.*

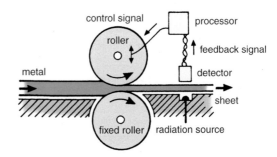

control signal

processor

roller

feedback signal

metal

detector

sheet

fixed roller radiation source

fermentation The process by which certain organisms such as **bacteria** and **yeasts** degrade organic compounds in the absence of **oxygen** in order to release **energy**. Fermentation is a form of anaerobic **respiration**. The following reaction shows fermentation by yeast (the basis of brewing and bread-making):

$$\text{glucose} \xrightarrow{\text{yeast}} \text{ethanol} + \text{carbon dioxide} + \text{energy}$$
$$C_6H_{12}O_6 \qquad 2C_2H_5OH \qquad 2CO_2$$

fertilization The fusing of haploid **gametes** during **sexual reproduction**. It results in the formation of a single cell called the *zygote* which contains the diploid number of **chromosomes**.

(a) *External fertilization*. This occurs when the gametes are passed out of the parents and fertilized, and development takes place outside the parents. External fertilization is common in aquatic organisms, such as fish and frogs, where movement of water helps the gametes to meet.

(b) *Internal fertilization*. This is particularly associated with terrestrial animals such as insects, birds and mammals. It involves the union of the gametes within the female's body. There are several advantages to internal fertilization:
   (i)   the **sperms** are not exposed to unfavourable dry conditions.
   (ii) the chances of fertilization occurring are increased.
   (iii) the fertilized **ovum** is protected within a shell (birds) or within the female body (mammals).

(c) *Fertilization in humans*. During the process of copulation the penis is inserted into the vagina and sperm **cells** (produced in the **testes**) are passed out of the penis. The sperms move through the **uterus** and into the oviducts. If an ovum is present in an oviduct, fertilization can occur there. The fertilized ovum (zygote) continues moving down the oviduct towards the uterus, dividing repeatedly as it does so. When it arrives at the uterus the zygote, which is now a ball of cells, becomes embedded

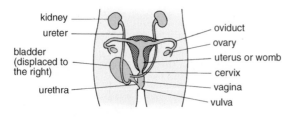

kidney
ureter
bladder (displaced to the right)
urethra

oviduct
ovary
uterus or womb
cervix
vagina
vulva

fertilization *Female reproductive organs.*

in the prepared wall of the uterus. This process is called implantation. All further development of the **embryo** occurs in the uterus.

fertilization *Male reproductive organs.*

(d) *Fertilization in plants.* During **pollination**, **pollen** grains are deposited on the stigmas of flowering plants. The pollen grains absorb nutrients and pollen tubes grow down through the style to the ovules in the **ovary**. A pollen grain enters an ovule through a small hole called a micropyle.

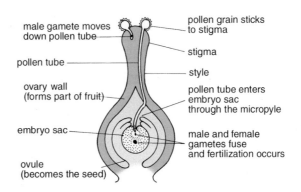

fertilization *Fertilization in plants.*

The bottom of each pollen tube breaks down and the male gamete enters the ovule and fuses with the female gamete. After fertilization the ovule, which contains the plant embryo, develops into a **seed** and the ovary develops into a **fruit**.

fertilizer A substance which is added to soil to increase the quantity or quality of plant growth.

When crops are harvested the natural recycling of soil **mineral salts** is disturbed; i.e. mineral salts absorbed by the crops are not returned to the soil. This is called *soil depletion* and may eventually render the soil infertile. Fertilizers replenish the soil. They are often described in terms of their nitrogen, phosphorus and potassium (NPK) content. Fertilizers may be in the form of organic fertilizers, such as sewage, or inorganic fertilizers, such as ammonium sulphate. Organic fertilizers often contain trace elements: these have to be added to inorganic fertilizers.

fetus or foetus The name given to a mammalian **embryo** after it develops its main features. In humans this is after about three months of **pregnancy**.

fibre optics The transmission of information along glass (or Perspex) fibres as pulses of **light**. The fibre is composed of two types of glass, one contained within the other. Light enters the fibre and travels along it as a result of *total internal reflection*.

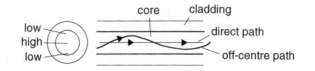

fibre optics *In a graded-index optical fibre the refractive index varies continuously. This bends light and keeps it inside the core.*

Optical fibres are very thin (less than 1 mm in diameter) and bend easily. They are far more efficient at carrying information than **copper** wires as very little **energy** is lost at each reflection. An optical fibre can carry far more information than a copper wire of similar diameter.

fission This occurs when the large unstable **nucleus** of an **atom** splits into two smaller stable nuclei of similar size, together with other smaller particles such as **neutrons**. The total **mass** of the products is less than that of the starting material. The difference in mass appears as **energy**. The fission of **uranium**-235 provides the energy in a **nuclear power** station.

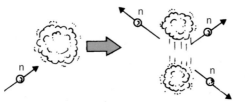

fission *The fission of a uranium-235 nucleus.*
$^{235}_{92}U + ^{1}_{0}n \rightarrow ^{236}_{92}U \rightarrow fission\ products + 3^{1}_{0}n + energy.$

fixing nitrogen Converting atmospheric nitrogen into **compounds** which are useful as **fertilizers**. Two important processes for fixing nitrogen are:
(a) The *Haber process* which converts atmospheric nitrogen into **ammonia**.
(b) The action of bacteria found on the roots of leguminous plants such as beans and peas.
See **denitrification, nitrogen** cycle.

flower This is the **organ** of **sexual reproduction** in flowering plants. The male part of the flower is called the **stamen** and the female part the **carpel**.

Sexual reproduction relies on the transfer of **pollen** from the stamen of one flower to the carpel of the same or a different flower.

fluid This describes a **liquid** or a **gas** where the particles can flow and are not fixed in a particular position, as is the case in a **solid**. Viscosity is a measure of the reluctance of a fluid to flow. It depends upon **temperature** because the particles' 'speed' depends on temperature. A viscous liquid like syrup flows better when it is warm.

food chain A relationship involving plants and animals in which **energy** and **carbon compounds**, made by green plants via **photosynthesis**, are passed to other living organisms, i.e. plants are eaten by animals which, in turn, are eaten by other animals:

$$
\begin{array}{llll}
\text{green plant} & \rightarrow \text{herbivore} & \rightarrow \text{small} & \rightarrow \text{large} \\
 & & \text{carnivore} & \text{carnivore} \\
\text{(producer)} & \rightarrow \text{(primary} & \rightarrow \text{(secondary} & \rightarrow \text{(tertiary} \\
 & \text{consumer)} & \text{consumer)} & \text{consumer)}
\end{array}
$$

The arrows indicate 'is eaten by'. There are many examples of food chains, e.g:

$$\text{grass} \rightarrow \text{sheep} \rightarrow \text{man}$$

Simple food chains like the one above seldom exist. More usually food chains are linked into other food chains, producing a *food web*.

All food webs are delicately balanced. If one link should be destroyed or the numbers of a particular organism should rapidly increase or decrease, all other organisms in the web will be affected. For example, in the pond food web, if the numbers of perch started to drop due to some disease the pike population would decrease (as they would have less food) while the water scorpion population would increase (as less of them would be eaten). The larger population of water scorpions would need more food, so there would be increased pressure on the populations of tadpoles and water beetles.

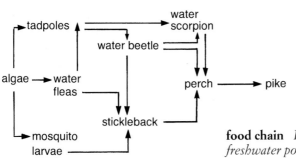

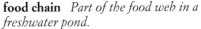

food chain *Part of the food web in a freshwater pond.*

force (F) A force changes an object's motion, making it move more or less quickly and/or making it change direction. The unit of force is the newton, N; one newton is the force which will accelerate one kilogram by one metre per second per second. Newton's laws of motion relate forces to their effects and lead to the expression:

$$\text{force} = \text{mass} \times \text{acceleration} \ (F = m \times a)$$

Here are some effects involving forces:

(a) *Gravity*. Attraction only – why objects fall to the ground when dropped.

(b) *Magnetism*. Attraction between unlike poles (N and S) and repulsion between like poles (N and N or S and S).

(c) *Electricity*. Attraction between opposite electrical charges (+ and –) and repulsion between similar charges (+ and + or – and –).

These forces act at a distance from the object they affect. Many forces act when there is, apparently, contact between particles experiencing the force. This occurs in **friction**, **weight**, push and pull (compressive and tensile) forces and in twisting. A *force meter* or *Newton meter* is sometimes used to measure force.

fossil The traces, or sometimes remains, of a plant or animal which have become embedded and preserved in rock.

fossil fuel Substances which have been formed over millions of years from the remains of animals and plants and which are used as fuels. **Coal**, **petroleum** and **natural gas** are examples of fossil fuels.

All fossil fuels are composed of compounds of the **elements carbon** and **hydrogen**. When they are burned (*see* **combustion**) in a plentiful supply of **air** they produce **carbon dioxide** and **water**. Fossil fuels also often contain small amounts of **compounds** containing **sulphur** and nitrogen which produce acidic gases when burned in air.

The world's reserves of fossil fuels are going down each year as, once used, they are not replaced by nature. Fossil fuels are nonrenewable energy **resources** (compare **renewable energy sources**). The ever-increasing amounts of fossil fuels burned around the world each year are causing some serious problems for our **environment**. *See* **greenhouse effect** and **acid rain**.

frequency (f) The number of cycles of a periodic motion in unit time. For example, the number of waves passing a point in one second or the number of times a pendulum swings to and fro in one minute. The **SI unit** is the hertz, Hz:

$$1 \text{ hertz} = 1 \text{ cycle per second}$$

The frequency, *wavelength* and **speed** of a **wave** are related to each other by the equation:

$$\text{speed} = \text{frequency} \times \text{wavelength} \ (c = f \times \lambda)$$

friction A **force** which tends to prevent motion between two surfaces. Sometimes friction works to our advantage. For example, friction prevents our feet slipping while walking, or a car skidding when driving. Unfortunately, friction is often a disadvantage. It slows down movement and wastes **energy**, resulting in **heat** and wear.

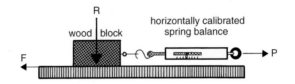

friction *Static and dynamic friction.*

If a block is placed on a bench and pulled gently, as shown above, at first nothing happens: the block doesn't move, thus, P = F. When P is gradually increased a maximum value is obtained before the block moves. This maximum value is the static or starting friction. Once the block starts to move, the value of P which is needed to keep it moving is called the dynamic or sliding friction. The dynamic friction is less than the static friction. If weights are put onto the block the value of the dynamic friction force increases in proportion. Even the smoothest surfaces are, in fact, rough when looked at with a microscope.

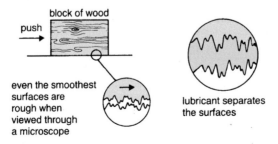

friction *(a) Even the smoothest surfaces are rough when viewed through a microscope. (b) Lubricant separates the surfaces.*

Frictionless motion is obviously an advantage in transport, since valuable **fuel** is used up overcoming friction. A hovercraft avoids friction with the ground by riding on a cushion of air.

fruit The ripened **ovary** of a **flower**. The fruit contains **seeds** formed as the result of **pollination** and **fertilization**. The fruit protects the seed and helps in its dispersal.

fruit and seed dispersal The methods by which most flowering plants spread their **seeds** far away from the parent plant. This is favourable to the plant because:
(a) It avoids competition for resources such as **water** and **light**.
(b) It ensures that the seeds will be spread over many different **habitats**, so there is a good chance that at least some of the seeds will flourish.
Most flowering plants disperse their seeds in one of the following ways:
(a) *Wind dispersal.* Air currents carry the fruits or seeds. These are usually light and are adapted to hang in the air as long as possible.
(b) *Animal dispersal.* This may happen in two different ways.

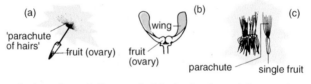

fruit and seed dispersal *Methods of wind dispersal*
(a) dandelion (b) sycamore (c) groundsel.

(i)   Some fruits, such as burdock, have hooks which stick to the coats of animals and may be brushed off some distance from the parent plant.
(ii) Berries such as the strawberry are succulent and are eaten by animals. The succulent part is digested but the small, hard fruits containing seeds pass through the alimentary canal of the animal undamaged. Eventually they are released in the faeces, often some distance from the parent plant.

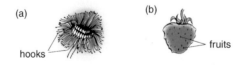

fruit and seed dispersal *Examples of fruit dispersed by animals*
(a) burdock (b) strawberry.

(c) *Explosive dispersal.* Some fruits, such as sweet pea, burst open with an effect rather like an explosion. This scatters their seeds away from the parent plant. This effect is the result of unequal drying of the fruit.

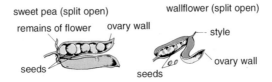

fruit and seed dispersal *Sweet pea and wallflower disperse their seeds by explosion of the fruit.*

fuel A substance which releases **heat energy** when it is treated in a particular way. In most fuels, energy is released by **combustion** (burning). Common fuels which produce energy by burning are paper, wood, **natural gas**, petrol and **coal**:

$$\text{fuel} + \text{oxygen (often from the air)} \rightarrow$$
$$\text{carbon dioxide} + \text{water} + \text{energy}$$

Nuclear fuels, like **uranium** and plutonium, produce energy when their **atoms** undergo nuclear **fission**.

fungus An organism which lacks chlorophyll, leaves and true roots and which lives as a **parasite** or a *saprophyte* (i.e. it feeds on dead matter). Examples are mushrooms, moulds and yeasts.

fuse A device in an electric **circuit** which melts and cuts off the supply of electricity if the **current** becomes too large. This might be due to a fault or an attempt to overload the system.

Nearly all conductors resist the flow of **charge**; the result is an **energy** transfer which causes the **temperature** of the conductor to increase. This is called the heating effect of a current.

Household fuses come in various ratings of amperes and are designed to melt if the current in the circuit rises above a particular value. To choose the correct fuse for an appliance which has a **power**, P, in **watts**, use the equation:

$$I = P/250$$

and select a fuse which has a value slightly higher than I. We now often use automatic switches called *circuit breakers* rather than fuses at currents higher than 13 amperes. The user can reset the switch after the fault is found and remedied.

The fuse protects the wires of the circuit and the appliance. It should be connected to the live wire. *See* **three-pin plug**.

fusion (nuclear fusion) The joining of two or more light atomic **nuclei** to make a more massive one whose **mass** is slightly lower than the combined masses of the particles it is made from. This process involves a large transfer of mass to **energy**. Much of the energy from the sun and other stars comes from fusion.

$$\,^2_1 H + \,^2_1 H \rightarrow \,^3_2 He + \,^1_0 n + energy$$

Fusion is used in an uncontrolled way in the hydrogen atom bomb. Effective ways of controlling this have not yet been developed; however, it promises to be a cheap and clean source of power in the future.

G

galaxy Millions of **stars** which appear as a group in the sky. Our **solar system** is within a galaxy which is called the *milky way*. The millions of stars in the milky way appear as a luminous band which encircles the heavens.

gamete A reproductive **cell** whose **nucleus** contains only half of the normal number of **chromosomes** (*haploid*), because these cells are produced by the process of double division called **meiosis**. In humans the male gametes are **spermatazoa** and the female gametes are **ova**.

   During **fertilization** male and female gametes fuse to form a *zygote*. The nucleus of the zygote contains the normal number of chromosomes (*diploid*).

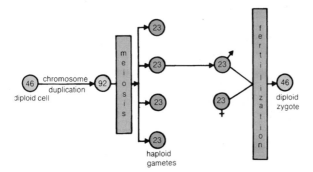

gamete *In humans, the haploid gametes have 23 chromo-somes and the diploid zygote, formed after fertilization has 46.*

gamma radiation (γ) High-energy (short wavelength) **electromagnetic waves** produced during the decay of certain **nuclei** (*see* **radioactivity**). **Energy** leaves as a packet of radiation (sometimes called a *quantum* or *photon*), and the nucleus becomes more stable.

   Gamma radiation can pass through matter very easily. It forms ions when it collides with **atoms** of the material and slowly loses energy. It is able to penetrate deep into the human body and may cause disorders such as cancers. Workers dealing with materials which emit gamma radiation must be protected by lead and/or concrete shielding. *Gamma radiography* is used to examine the internal structure of metal objects such as aircraft engine components. Faults which are not visible in an exterior examination can be detected and the component replaced before the engine fails.

gas or vapour The normal **state of matter** whose particles attraction which hold the particles together in have the highest **energy**. When **heat** is supplied to a **liquid** the **atoms** or **molecules** are given **kinetic energy**. This may be enough to overcome the forces of the liquid state. If this happens the liquid boils and turns to gas. Collisions of gas particles on the walls of a container exert a **pressure**. The particles move quickly, at hundreds of metres per second, in random directions and fill all the space available. The **density** of a gas is much less (e.g. 1/1000 times) than a **solid** or a liquid. Natural gas is a major energy **resource**.

gas exchange In this process organisms exchange gases with the **environment** for the purposes of **metabolism**. Most organisms need a continuous supply of **oxygen** for the energy-producing reaction of **respiration**:

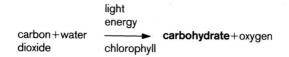

In addition to this green plants need carbon dioxide for **photosynthesis**:

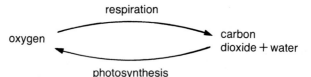

Both reactions use and produce gases which are exchanged between the organism and the atmosphere (land organisms) or water (aquatic organisms).

respiration

oxygen   carbon dioxide + water

photosynthesis

gas exchange *Both respiration and photosynthesis involve gas exhange with the environment.*

The gas-exchange surfaces within organisms have certain characteristics which allow the exchange to take place:
(a) A large surface area for maximum gas exchange.
(b) The surface is thin to allow **diffusion**.
(c) The surface is moist as gas exchange takes place in solution.
(d) In animals, the surface has a good **blood** supply as it is the blood which transports gases to and from the cells of the body.
In mammals gas exchange occurs across the alveoli in the **lungs**. It is the result of differences between the concentrations of oxygen and carbon dioxide in the air inside the alveoli and the deoxygenated blood in the

capillaries around the alveoli. These differences in concentration are called *concentration gradients*. These gradients cause oxygen to diffuse from the alveoli into **red blood cells** in the capillary, and carbon dioxide to diffuse out of the blood into the alveoli. *See* **breathing**.

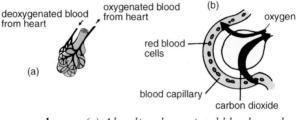

gas exchange *(a) Alveoli and associated blood vessels. (b) Gas exchange in the alveolus.*

general formula A formula used in organic chemistry which shows the relative numbers of the different **atoms** in terms of a variable n for all of the members of a group of **compounds**. The actual formula for any particular compound in the group is found by substituting a number for n. For example; the general formula for the group of compounds called alkanes is C_nH_{2n+2}. The formula for each member of the alkanes is obtained by substituting for n.

A series of compounds with the same general formula is sometimes referred to as a *homologous series*. *See* **hydrocarbons**.

n	Name of compound	Molecular formula of compound
1	methane	CH_4
2	ethane	C_2H_6
3	propane	C_3H_8
4	butane	C_4H_{10}
etc.		

generator A machine which converts **kinetic energy** to **electricity**. *Alternators* give **alternating current**; other types of generator give **direct current**. Simple forms of both AC and DC generators have a similar structure. *See* **electromagnetic induction**.

genes The subunits of **chromosomes**. They consist of lengths of **DNA** and control the hereditary characteristics of organisms. A single gene consists of up to 1000 *base pairs* in a DNA **molecule**. The sequence of the base pairs represents coded information known as the *genetic code*. This code determines the structures of the different types of **proteins**, particularly **enzymes**, synthesized by the cell. In turn, these determine the structure and function of the cells and **tissues**, and ultimately the organism.

The genetic code is an arrangement of *nitrogen base pairs* in DNA. Each group of three adjacent base pairs (called triplets) is responsible within the cell for linking together **amino acids** to form protein. For example, the base triplet GTA codes for the amino acid *histidine* while GTT codes for another amino acid, *glutamine*.

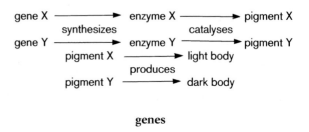

The sequence and type of amino acids determine the nature of the protein. Consider two fruit flies (*Drosophila*): one has a light body colour controlled by a gene X, and the other a dark body colour controlled by a gene Y.

genetic engineering The deliberate alteration of the genetic information on the **chromosomes** of an organism in order to bring about some advantage to people. This technique has been used successfully to produce microorganisms which can synthesize useful drugs, such as **insulin**. Controversially, it is now possible to improve the food potential of animals and plants by engineering qualities such as high yields and resistance to disease.

genotype *See* **monohybrid inheritance**.

geothermal aquifer A source of hot water from the ground which may be used for heating. In some parts of the world there are very hot rocks quite close to the surface. As rainwater percolates through these rocks it is heated. If the geology of the location is suitable, the hot water collects in deposits which can be tapped off by drilling down into the rocks. *See* **renewable energy sources**.

geothermal hot dry rock structures Rock structures from which heat energy can be removed. The rocks are hot but have no water associated with them (*see* **geothermal aquifer**). Two wells are drilled down to the rocks, ending several hundred metres apart. The permeability of the rock between the ends of the wells is increased by applying hydraulic **pressure** which enlarges natural fissures in the rock.

Cold water is forced down one well and passes through the fissures in the hot rocks between the ends of the wells. As it passes through the fissures it is heated and hot water is collected from the second well. *See* **renewable energy sources**.

germination The first stage in the **growth** of **spores** and **seeds**. This often follows a period of dormancy and normally requires particular environmental conditions such as the availability of **water** and **oxygen** and a favourable **temperature**. If these conditions are not present, spores and seeds can remain alive for some time (often years) before germinating. In this state they are said to be *dormant*.

Seed germination in flowering plants. There are two types of germination; hypogeal and epigeal. The difference lies in what happens to the cotyledons. In epigeal germination the cotyledons become the first leaves of the new plant, while in hypogeal germination they do not.

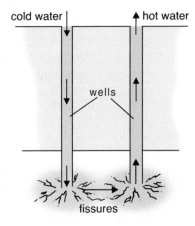

cold water | hot water

wells

fissures

geothermal hot dry rock structures

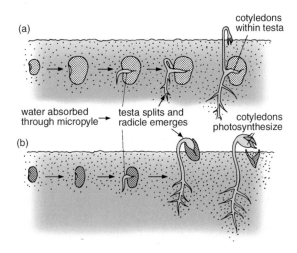

(a)

cotyledons within testa

water absorbed through micropyle → testa splits and radicle emerges

cotyledons photosynthesize

(b)

germination *(a) Hypogeal germination (broad bean). (b) Epigeal germination (French bean).*

In both cases water is absorbed through the micropyle, the testa splits open and the radicle emerges.

gestation period *See* **pregnancy**.

giant structure Structures of **atoms** or **ions** in which there are large numbers of particles present in a crystal lattice. Each particle has a strong force of attraction for the particles around it. Giant structures tend to have high **melting** and **boiling points**. **Ionic compounds**, **metals** and some **nonmetals** have giant structures. Compare **molecule**.

glucose A monosaccharide **carbohydrate** made during **photosynthesis**. This substance is an important **energy** source in both animal and plant **cells**. *See* **respiration**.

glycogen This is a polysaccharide **carbohydrate**. It consists of branched chains of **glucose** units. It is important in animals as a short-term energy store. In vertebrates glycogen is stored in **muscle** and **liver cells**. It can be rapidly converted to glucose by the action of amylase **enzymes** if the level of glucose in the blood is too low. *See* **insulin**.

gravitational potential energy The potential energy that an object has by virtue of its position about the ground. It depends upon the **mass** of the object, its height above the ground and the gravitational field strength:

Gravitational potential energy (J) = mass (kg) × gravitational field strength (N/kg) × vertical height (m).

On the **Earth** the gravitational field strength is 9.8 N/kg but the value is often taken as 10 N/kg for ease of calculation.

gravity A **force**, the attraction between **masses**. The cause of gravity is not known. The **weight** of an object depends on the value of the gravitational attraction and its mass. Gravity is taken to act through a point at the centre of an object where we imagine the object's mass to be concentrated. This point is sometimes called the *centre of gravity* or *centre of mass*.

greenhouse effect The effect created by **heat radiation** from **Earth** being reflected back to Earth by the **atmosphere**. **Temperatures** around the world have shown a small but significant increase over the last century. Some scientists believe that this is linked to the increasing concentration of **carbon dioxide** in the atmosphere. The Earth receives heat radiation from the Sun. Some of this is absorbed by the Earth and re-emitted back into space at a longer wavelength. The atmosphere reflects some of the

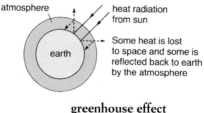

greenhouse effect

re-emitted radiation back on the Earth and prevents it from going into space. The more carbon dioxide there is in the atmosphere the more heat radiation is reflected back to Earth, hence the hotter the Earth will become. A similar effect happens in a greenhouse; the glass reflects re-emitted heat radiation back into the greenhouse (hence the name of this effect).

The increasing concentration of carbon dioxide in the atmosphere is a result of the combustion of increasing quantities of **fossil fuels** and the large scale deforestation in many parts of the world (less vegetation, therefore less **photosynthesis**). The greenhouse effect is a serious problem because it is believed to be responsible for changes in the climate of some areas of the world and it may result in an increase in the sea level as part of the ice caps melt.

group In the Periodic Table the **elements** are arranged in horizontal periods and vertical groups:

Group I	(alkali metals) lithium, sodium, potassium
Group II	(alkali earth metals) beryllium, magnesium, calcium
Group III	boron, aluminium
Group IV	carbon, silicon
Group V	nitrogen, phosphorus
Group VI	oxygen, sulphur
Group VII	(halogens) fluorine, chlorine, bromine
Group 0	(noble gases) helium, neon, argon

In each group the outermost electron shell contains the same number of **electrons** for each member. The number of electrons is the same as the group number.

growth The increase in size and complexity of an organism during its development from **embryo** to maturity. It is the result of **cell** division, cell elongation and cell differentiation. In plants growth originates at certain localized areas called *meristems*. In animals growth goes on all over the body.

H

habitat The place where an animal or plant lives. Organisms are adapted to the particular environmental conditions within a habitat.

half-life period ($T_{1/2}$) The time it takes for a measure, whose **decay** is exponential, to fall to half of its value. Half-life is often used in connection with the number of parent radioactive **nuclei** in a sample.

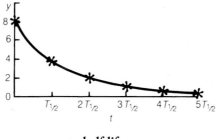

half-life

halogens A collective name for the **elements** in **Group VII** of the **Periodic Table**. They are all poisonous nonmetallic elements. At room temperature fluorine and **chlorine** are gases, bromine is a liquid and iodine is a solid. All the halogens are oxidizing agents; their oxidizing power and chemical reactivity decreases as we go down the group, i.e:

$$F_2 > Cl_2 > Br_2 > I_2$$

They react vigorously with **metals** and **hydrogen** forming *halides*:

fluorine → fluorides
chlorine → chlorides
bromine → bromides
iodine   → iodides

They all contain seven **electrons** in the outer shell of the **atom** and form **ions** with a single negative charge (*univalent anions*) e.g: Cl⁻, Br⁻.

heart A muscular pumping organ which maintains the circulation of **blood** around the body. It usually contains valves which prevent the blood from flowing backwards. In mammals the heart has four chambers; two relatively thin-walled atria (auricles) which receive blood, and two thicker-walled ventricles which pump blood out. The right side of the heart deals only with deoxygenated blood and the left side only with oxygenated blood. The wall of the left ventricle is thicker and more powerful than that of the right since it pumps blood to all parts of the body, whereas the right ventricle only pumps blood to the **lungs**. *See* **circulatory system.**

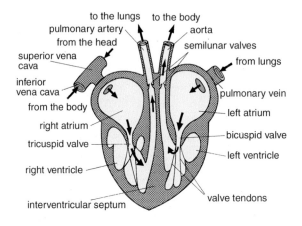

heart *Structure of the mammalian heart.*

heartbeat This is the result of alternative contraction and relaxation of the **heart**. In mammals it consists of two distinct phases:

(a) *Diastole.* The atria and ventricles relax. This allows **blood** to flow from the atria into the ventricles.

(b) Systole. The ventricles contract, forcing blood into the pulmonary **artery** and the aorta. At the same time the relaxed atria fill with blood ready for the next beat.

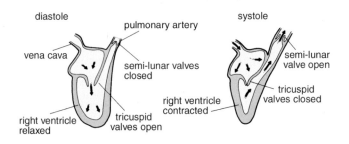

heartbeat *Relaxation and contraction of heart (right side).*

Heartbeat is initiated by a structure in the right atrium called the *pacemaker*, although the rate is controlled by a part of the **brain** called the medulla oblongata. This part of the brain monitors the activity of the body by detecting increases in the **carbon dioxide** level in the blood; the result of increased **respiration** (caused by increased activity). It is also affected by certain **hormones** such as adrenaline from the adrenal glands. The rate of human heartbeat is measured by counting the **pulse rate**. People with heart defects are sometimes fitted with electronic pacemakers.

heat A form of **energy**. It is the internal energy possessed as the movement of **molecules** of a substance (*see* **kinetic energy**). The word is also used to mean energy transfer. Thus when we heat a beaker of water with a bunsen burner there is energy transfer from the burning **gas** to increase the particle motion in the **water**. Heat transfer may be by **conduction** (thermal), **convection** or **radiation**.

homeostasis A general term for the maintenance of constant conditions within an organism. For example:
(a) control of **blood glucose** level by **insulin**.
(b) control of blood water content by ADH.
(c) control of body **temperature** by the **skin**, etc. *See* **hormones**.

Hooke's law The law which states that when an elastic object, like a rubber band, is stretched, the increase in its length is proportional to the **force** applied; provided this force does not exceed the *elastic limit*. For example, if a force of 10 N increased the length of a rubber band by 1 cm then a force of 20 N would result in an increase of 2 cm. If the elastic limit is exceeded the object ceases to be elastic and no longer obeys Hooke's law. *See* **elasticity**.

hormones Chemicals which are secreted by the **endocrine glands** and are transported via the bloodstream to certain target **organs**. At these organs the hormones cause specific effects which are vital in regulating and coordinating the activities of the body. Hormones are sometimes called *chemical messengers*. Hormone action is usually slower than nervous stimulation. The following tables summarize the properties of some important human hormones, but there are many others. (*See* table on pages 71–72)

Endocrine gland	Hormone	Effects
Pituitary gland	ADH (anti-diuretic hormone)	Controls water reabsorption by the kidneys.
	TSH (thyroid stimulating hormone	Stimulates thyroxine production in the thyroid gland.
	FSH (follicle-stimulating hormone	Causes ova to mature and the ovaries to produce oestrogen.
	LH (luteinizing hormone)	Initiates ovulation and causes the ovaries to release progesterone.
	Growth hormone	Stimulates growth in young animals. In humans, deficiency causes dwarfism and excess causes gigantism.
Thyroid gland	Thyroxin	Controls rate of growth and development in young animals. In human infants, deficiency causes cretinism. Controls the rate of chemical activity in adults. Excess causes thinness and over-activity, and deficiency causes obesity and sluggishness.
Pancreas (Islets of Langerhans	Insulin	Stimulates conversion of glucose to glycogen in the liver. Deficiency causes diabetes.
Adrenal glands	Adrenalin	Under conditions of 'fight, flight or fright' causes changes which increase the efficiency of the animal.

Endocrine gland	Hormone	Effects
		For example, increased heartbeat and breathing, diversion of blood from gut to muscles, coversion of glycogen in the liver to glucose.
Ovaries	Oestrogen	Stimulates secondary sexual characteristics in the female, for example, breast development. Causes the uterus wall to thicken during menstrual cycle.
	Progesterone	Prepares the uterus for implantation.
Testes	Testosterone	Stimulates secondary sexual characteristics in the male, for example, facial hair.

human genome project A venture costing in excess of US $6 billion and involving over 1000 scientists from 50 countries around the world that will be completed in Spring 2003. The objective is to trace every single human gene (over 30,000) and find its position on a chromosome. In the future this will help scientists to understand how certain diseases affect people and they can best be treated.

humus A general term for organic material in **soil**, consisting of decomposing plant and animal remains. It is a desirable feature of soils as it provides plants, and ultimately animals, with nutrients.

hybrid A plant or animal produced as a result of a cross between two parents who are genetically unlike each other, or between two differing but related organisms.

hydrocarbon A **compound** which contains only **hydrogen** and **carbon**. There are three main series of hydrocarbons.

Name of series	General formula	Examples	
Alkanes	C_nH_{2n+2}	methane	CH_4
		ethyane	C_2H_6
		propane	C_3H_8
Alkenes	C_nH_{2n}	ethene	C_2H_4
		propene	C_3H_6
Alkynes	C_nH_{2n-2}	ethyne	C_2H_2
		propyne	C_3H_4

hydrochloric acid A strong acid made by dissolving *hydrogen chloride* (HCl) gas in water.

Hydrogen chloride gas can be made by reacting the **elements** together directly or by reacting sodium chloride with concentrated **sulphuric acid**:

$$H_2(g) + Cl_2(g) \rightarrow 2HCl(g)$$
$$NaCl(s) + H_2SO_4(l) \rightarrow NaHSO_4(s) + HCl(g)$$

The gas reacts with **ammonia** to form dense white fumes of ammonium chloride:

$$NH_3(g) + HCl(g) \rightarrow NH_4Cl(s)$$

The **atoms** in hydrogen chloride gas are held together by **covalent bonds**; however, **ions** are formed when it dissolves in a *polar* solvent such as water, hence hydrochloric acid is an ionic compound and contains hydrogen (H^+) and chloride (Cl^-) ions.

Hydrogen chloride gas is very soluble in water. The maximum **concentration** of the solution is 36% (about 11 **moles/dm³**). The acid is *monobasic* and produces **salts** called chlorides.

$$Fe(s) + 2HCl(aq) \rightarrow FeCl_2(aq) + H_2(g)$$
$$Mg(s) + 2HCl(aq) \rightarrow MgCl_2(aq) + H_2(g)$$

The acid releases carbon dioxide from carbonates and can be oxidized to **chlorine**.

$$Na_2CO_3(s) + 2HCl(aq) \rightarrow 2NaCl(aq) + CO_2(g) + H_2O(l)$$
$$MnO_2(s) + 4HCl(aq) \rightarrow MnCl_2(aq) + 2H_2O(l) + Cl_2(g)$$

hydroelectricity Electrical power produced by using the **energy** of falling **water**, often from behind a dam.

hydrogen (H_2) A reactive **element** which is a diatomic **gas**. An **atom** of the commonest isotope of hydrogen consists of a **nucleus** of one **proton** and, outside the nucleus, one orbiting **electron**. Hydrogen has two other

isotopes, *deuterium* and *tritium*, whose nuclei also contain one and two **neutrons** respectively. Hydrogen can form **covalent bonds** by sharing electrons; e.g:

$$2H_2(g) + O_2(g) \rightarrow 2H_2O(g)$$
$$C_2H_4(g) + H_2(g) \rightarrow C_2H_6(g) \text{ (}hydrogenation\text{)}$$

Hydrogen ions (H^+) are produced when hydrogen atoms lose their electrons. (Under the right conditions a hydrogen atom can produce hydride ions, H^-). **Acids** contain hydrogen ions. Hydrogen is a reducing agent. It is used to convert vegetable oils into *margarine*.

Industrially, hydrogen is made from **petroleum** by the *steam reformation of alkanes*. Large quantities of hydrogen are used in the formation of **ammonia** by the *Haber process*. In the laboratory hydrogen is made by reacting a **metal** with an acid other than nitric acid. Zinc and dilute **hydrochloric acid** are commonly used:

$$Zn(s) + 2HCl(aq) \rightarrow ZnCl_2(aq) + H_2(g)$$

Hydrogen is also released by the **electrolysis** of aqueous solutions which contain ions of elements above hydrogen in the **electrochemical series**, e.g. $NaCl(aq)$, $Mg(NO_3)_2(aq)$, and by the reaction of **water** with **Group I** and Group II metals, e.g. **sodium**, **calcium**.

Hydrogen is highly inflammable and forms explosive mixtures with **oxygen**. It has a very low **density**, and was once used in balloons and airships, however several large-scale disasters occurred due to its highly inflammable nature.

hydrogen ion (H^+) A positively charged **ion** (which is, in fact, a **proton**) formed when a hydrogen atom loses its **electron**:

$$H \rightarrow H^+ + e^-$$

This small particle is very reactive. In aqueous solution it combines with a **molecule** of **water** to form an *oxonium ion*:

$$H^+ + H_2O \rightarrow H_3O^+$$

All aqueous solutions of **acids** contain more hydrogen ions than **hydroxide ions**. The **pH** of a solution is a measure of the **concentration** of hydrogen ions in it.

hydroxide (OH^-) Found in all **alkalis** and in alkaline solutions, e.g. sodium hydroxide Na^+OH^-. It is present, to a small extent, in all aqueous solutions because of the **dissociation of water**:

$$H_2O \quad \rightarrow \quad H^+ \quad + \quad OH^-$$
water   hydroxide   hydroxide ion

Solutions which contain more hydroxide ions than hydrogen ions are described as alkaline. These have a **pH** greater than 7.

I

igneous rocks Rocks formed by the crystallization of molten material, called *magma*, which comes up from deep in the earth. Igneous rocks are classified according to their grain size and composition. Composition is often thought of as ranging from acidic to basic because of the nature of the different minerals present.

	Acidic		**Basic**	
Coarse grain	Granite	Diorite	Gabbro	Peridotite
Fine grain		Felsite		Basalt

immunization The introduction of **antigens** into the body in the form of a **vaccine**. This produces an *immune response* to that antigen, thus protecting the person against future attack by the same type of antigen. In Britain, all parents have the opportunity to have their children immunized against a variety of diseases. Babies are normally immunized against diphtheria, tetanus, whooping cough, polio, measles, mumps and rubella (German measles). All teenagers are immunized against tuberculosis.

incubator A device for maintaining a constant **temperature** suitable for the **growth** and development of bacterial cultures. The temperature is often set at around human body temperature, 36.9 °C. Incubators are also used for hatching eggs.

indicator A substance which turns different colours according to acidic and alkaline conditions. It can be used to show whether a solution is an **acid** or an **alkali**:

Indicator	Colour in acid	Neutral	Colour in alkali
litmus	red		Blue
phenophthalein	colourless		Red
universal	red orange yellow	green	blue purple

infrared A region of the **spectrum** of **electromagnetic waves**. The approximate wavelength range is 10^{-6}–10^{-3} m, and the approximate **frequency** range 10^{12}–10^{15} Hz.

Infrared **radiation** is thermal or heat radiation. It is detectable by blackened **thermometers**, **skin nerve cells**, thermopiles and photographic films. All matter radiates infrared at all times.

inherited diseases Diseases resulting from **genes** which produce harmful effects. These genes may be dominant or recessive. When the disease is caused by a recessive gene, carriers may transfer the disease to the next generation without, themselves, suffering from its effects. *Cystic fibrosis* is a serious defect of the **pancreas** caused by a recessive gene. Some inherited diseases, such as the blood disease *haemophilia*, are associated with genes on the sex **chromosomes**.

inorganic chemistry A branch of chemistry concerned with the study of those aspects which do not fall within the study of **organic chemistry**. It is the study of **elements** and their **compounds**. This includes the chemistry of some compounds of the element **carbon**, such as its oxides, metal carbonates and hydrogencarbonates, but excludes all organic compounds such as alcohols, esters and **hydrocarbons**.

insulation A technique to reduce the **energy** transfer from one place to another. Thus *thermal insulation*:
(a) Reduces **heat** energy losses from a high temperature area such as an oven.
(b) Reduces heat energy gains to a low temperature area such as a refrigerator.
   In cold climates people use many methods of insulation to reduce heat loss from their homes. These may include *double glazing, cavity wall insulation and loft insulation*, making use of such materials as *fibre-glass, polystyrene and mineral wool*. Electrical insulation often involves coating conductors with **nonmetals**, such as plastics, to prevent **charge** (and thus energy) being transferred anywhere other than along the conductor.
   Metals are good conductors of both thermal and electrical energy. Their value in insulation is limited to use as reflectors of **radiation**.

insulin A **hormone** which is secreted by the Islet of Langerhans in the **pancreas** of vertebrates.
   Insulin regulates the conversion of **glucose** into **glycogen** which occurs in the **liver**. If the concentration of glucose in the **blood** is high, the rate of secretion of insulin is high. Thus glucose is rapidly converted into glycogen for storage in the liver. Conversely, if the concentration of glucose in the blood is low, less insulin is secreted, and less glucose is converted into glycogen. This is an example of the **feedback** regulation associated with many hormones. People suffering from *diabetes* are unable to produce enough insulin to control the delicate glucose level balance in their bodies.

insulin *Feedback regulation of insulin secretion.*

intestine This is the region of the **alimentary canal** between the stomach and the anus or cloaca. In vertebrates, most **digestion** and **absorption** of food occur in the intestine which is usually differentiated into the small intestine and the large intestine.

ion An **atom** which has become electrically charged by gaining or losing **electrons**. *Cations* are positively charged, e.g. Na^+, and *anions* are negatively charged, e.g. O^{2-}. Atoms tend to lose or gain electrons to produce an ion with the same *electronic configurations* as a **noble gas**. Groups of atoms (radicals) may also form ions:

$$\text{sulphate } SO_4^{2-} \qquad \text{nitrate } NO_3^-$$
$$\text{hydroxide } OH^- \qquad \text{ammonium } NH_4^+$$

ionic bonds Chemical bonds which occur because of *electrostatic* attractive forces between positively and negatively charged **ions**. Ionic bonds are often present in **compounds** of **nonmetals** from **Groups** VI and VII and **metals**, e.g. $Na^+ Cl^-$, $Mg^{2+}Br_2^-$ and in compounds containing, *radicals* such as sulphate and nitrate, e.g. $Cu^{2+} SO_4^{2-}$, $K^+ NO_3^-$.

ionic compounds Compounds which contain **ionic bonds**. They tend to have high **melting** and **boiling points** and are conductors of electricity when molten.

ionizing radiations **Radiations** which can remove **electrons** from **atoms**, thus creating **ions**. This may be caused by **alpha** and **beta particles**, gamma rays (*see* **gamma radiation**) and **X-rays** (**neutrons** cause ionization indirectly).

iron The most widely used metallic **element**, although it is usually encountered as **steel alloys**. In this form it is used for building girders, car bodies, cans, tools and many other items of everyday life.

  Iron is extracted from **ores** such as haematite (Fe_2O_3) and magnetite (Fe_3O_4) by reduction with carbon monoxide in a blast furnace. The iron produced in a **blast furnace** is brittle. To make it useful it is converted into a range of different steels with varying properties.

A major problem in using iron and many steels is **corrosion** (**rust**). The iron oxidizes in moist **air** to produce a soft crumbly oxide:

$$\text{iron} \xrightarrow{\text{moist air}} \text{iron(II) oxide } (Fe_2O_3)$$

Iron is a transition metal and can have **valency** 2 or 3. It reacts with dilute **acids** to form iron(II) **compounds**, for example:

$$Fe(s) + 2HCl(aq) \rightarrow FeCl_2(aq) + H_2(g)$$

Small amounts of iron (in the form of iron compounds) are an essential part of the human diet. It is needed by the body to make the pigment haemoglobin which gives the red colour to **red blood cells**.

isotopes Species of the same chemical **element**. They have the same number of **protons** but different numbers of **neutrons** in their **nuclei**. The following table shows details of some isotopes of **carbon**.

	Number of protons	**Number of neutrons**
carbon-12	6	6
carbon-13	6	7
carbon-14	6	8

J

joint The point in a **skeleton** where two or more **bones** meet. Some joints allow movement. There are three main types of movable joint in mammals.
(a) *Ball and socket joint.* This allows movement in several planes. In humans the shoulder and hip are ball and socket joints.

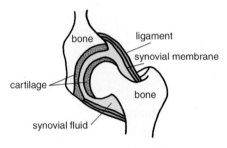

joint *Ball and socket joint.*

(b) *Hinge joint.* This allows movement in only one plane. In humans the elbow and knee are hinge joints.

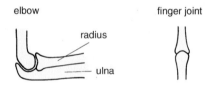

joint *Hinge joint.*

(c) *Gliding joint.* This type of joint occurs when two flat surfaces are able to glide over each other, allowing only a small amount of movement. In humans the wrist and ankle are gliding joints.

joint *Gliding joint.*

The ends of bones at movable joints are covered in *cartilage* to reduce **friction** as the bones rub against each other.

joule (J) The unity of **energy** and work. 1 joule of work is done when a **force** of 1 newton moves 1 metre.

K

Kelvin scale A scale of **temperature** sometimes used in preference to the Celsius (or centigrade) scale. The Celsius scale is positive above the **melting point** of ice (0 °C) and negative below it. Celsius is a widely used scale; however, in some fields of study, such as low temperature physics, negative values are a disadvantage. The Kelvin (or absolute) scale does not have negative values since its zero is at *absolute zero*. To convert Celsius temperature to Kelvin temperature add 273, i.e.:

$$0 \text{ °C} = 273 \text{ K}$$
$$100 \text{ °C} = 373 \text{ K}$$

kidney An **organ** found in vertebrates which consists of a collection of units called *nephrons*. It is concerned with **excretion** and osmoregulation. In humans, the kidneys are a pair of red-brown oval structures at the back of the abdomen.

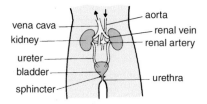

kidney *The position of the kidneys and associated organs in the human body.*

Oxygenated **blood** enters each kidney via the renal **artery** and deoxygenated blood leaves by the renal **vein**. A tube called the *ureter* connects each kidney with the bladder.

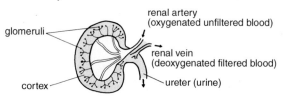

kidney *Section through a kidney.*

Within the kidney the renal artery divides into many small blood vessels called *arterioles* which terminate in tiny knots of blood **capillaries** called *glomeruli*. There are about one million glomeruli in a human kidney. Each is enclosed in a cup-shaped organ called a *Bowman's capsule*.

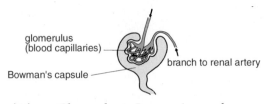

kidney *Glomerulus in Bowman's capsule.*

Two processes occur in the kidneys:

(a) *Ultrafiltration.* The blood vessel leaving the glomerulus is narrower than the one entering it. This puts the blood within the glomerulus under high **pressure**. Under these conditions the capillary wall acts as a selectively permeable membrane. Blood components which have only small molecules are forced through it into the Bowman's capsule while components with larger molecules remain in the blood.

Large particles (unfiltered)	Small particles (filtered)
Blood cells	Glucose
Plasma proteins	Urea
	Mineral salts
	Water
	Amino acids

(b) *Reabsorption.* The fluid filtered from the blood passes from the Bowman's capsule down the renal tubule. Each tubule is supplied with a network of blood vessels and as the fluid passes through it useful materials are reabsorbed into the blood. These will include all the glucose and amino acids and some mineral salts and water.

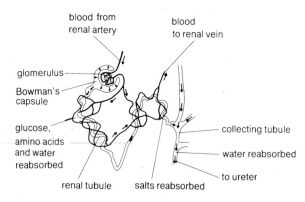

kidney *Reabsorption of useful substances.*

By the time the liquid reaches the end of the renal tubule it consists of a solution of urea and mineral salts. It drains into the bladder via the ureter and is eventually expelled from the body as **urine**. The amount of water which is reabsorbed is controlled by a **hormone**, called the anti-diuretic hormone (ADH). It is released by the pituitary gland, which monitors the amount of water in the blood. If the level of water is too low more ADH is released into the bloodstream. This results in more water being reabsorbed by the kidneys, leaving the urine more concentrated. If the level of water is too high the converse occurs.

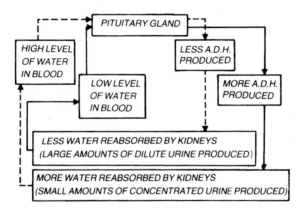

antidiuretic hormone *Control of water loss.*

kinetic energy The type of **energy** existing in a moving object. The amount depends on the object's **mass** m and its **speed** v as follows;

$$\text{kinetic energy} = \tfrac{1}{2} \text{ mass} \times \text{speed}^2$$
$$W = \tfrac{1}{2}mv^2$$

kinetic model A theory about the structure of matter which is commonly accepted by scientists. It is described by the following four statements:
(a) All matter is made of particles.
(b) The particles are always moving.
(c) The particles attract each other.
(d) The **temperature** of a substance relates to the mean energy of its particles.

The kinetic theory is able to explain the behaviour of matter. *See* **states of matter**.

L

landfill gas An inflammable mixture of gases obtained when certain waste materials are buried in the ground and allowed to decompose by bacterial action. This **renewable energy source** is a significant source of **energy** in the USA but has been less well exploited in Britain to date.

latent heat Hidden **heat**. It is the **energy** involved in changes of **state**. If heat is added, at a constant rate, to a **solid**, the **temperature**/time graph has shape (a). A similar graph for a **liquid** gradually losing heat at a constant rate has shape (b).

In each case, the temperature stays constant while the change of state takes place. A similar situation exists in the changes from liquid to **gas** and gas to liquid. The quantity of energy transferred to or from the particles during changes of state depends on the nature of the substance and its state.

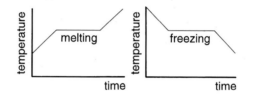

latent heat *Temperature remains constant during state changes between (a) solid to liquid; (b) liquid to solid.*

lead A metallic **element** in **Group** IV of the **Periodic Table**. It is a soft grey **metal** and has a high **density**. The main source of lead is the ore galena (PbS). Lead and its **alloys** are widely used, e.g. in the lead-acid accumulator solders, and as flashing on roofs to keep out water. Lead absorbs **radiation** from radioactive **isotopes** and **X-rays**. It is used for screening and containing radioactive sources to protect people from the effects of radiation. Lead can be used to store highly corrosive chemicals such as **sulphuric acid**.

At one time lead was widely used for water pipes, and some of its **compounds** were the basis of paints; however, studies have shown that even very small amounts of lead absorbed into the body can damage a person's health. A lead compound, tetraethyl lead, is still used as an antiknock agent in petrol (making the engine run more smoothly), however, because of the danger to health, many car manufacturers are designing cars which can use other antiknock agents.

leaf The part of a flowering plant which grows from the **stem**. It is typically flat and green, and contains **chlorophyll**. The functions of a leaf are:

(a) **Photosynthesis**.
(b) **Gas exchange**.
(c) **Transpiration**.

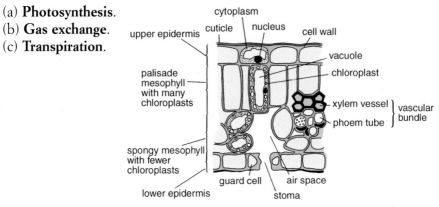

leaf *Structure of a dicotyledon leaf.*

LED Light-emitting diode. This is a p-n junction **diode**, and is usually made from gallium arsenide phosphide. **Energy** is released within the LED and this is given off as **light**. The junction is made near to the surface so that the emitted light can be seen. No light is emitted with a reverse bias. LEDs are commonly coloured red, yellow or green. They are widely used in a variety of electronic devices.

lens An optical device which bends transmitted **light** by **refraction**.
(a) *Converging (convex) lenses* bring the rays together
    at a point called the *focus*. They are thicker at the centre than at
    the edge.

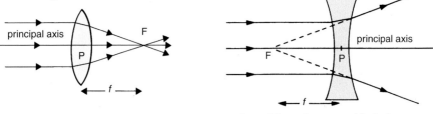

lens *The refraction of light by   **lens** *The refraction of light by
      a converging lens.*   a diverging lens.*

(b) *Diverging (concave) lenses* spread the rays out so they appear to be
    coming from one point, the focus. They are thin at the centre but get
    thicker towards the outside. *See* **eye**.

leucocyte *See* **white blood cell**.

lever A type of machine in which a certain **force** applied at one point gives an output force elsewhere. Levers are often grouped into three classes (1, 2 and 3). The classes differ in the positions of the input force (the effort applied) and the output force (the load overcome) relative to the pivot (turning point or fulcrum).

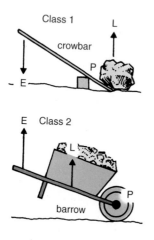

The pivot is between the load and the effort. Usually the effort is smaller than the load because it is further out.

The pivot is at the end of the lever and the load is in the middle. A small effort lifts a large load.

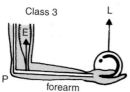

The pivot is again at the end but the effort is in the middle. There is a mechanical disadvantage and a large effort is needed to lift a small load

lever *The main types of lever and how they operate.*

light A region of the **spectrum** of **electromagnetic waves** which can be seen by the normal human eye. It has wavelengths between 400–760 nm. The colour of the light is related to its wavelength. Like all waves, light can be absorbed, reflected, refracted, diffracted and shows interference effects. Like all electromagnetic waves, it travels through empty space at 300 000 000 m/s and this value is commonly called the **speed of light**.

lipids *See* **fats**.

liquid A **state of matter** in which particles are bonded by inter-molecular forces.

The particles are not fixed in a rigid lattice as in a **solid**. During **evaporation** and boiling the particles get enough energy to overcome the intermolecular forces and become a **gas**.

litmus A dye extracted from lichen. It is used as an acid–base **indicator**.

acidic solution	pH 7	alkaline solution
red	purple	blue

liver The largest **organ** of the vertebrate body which occupies much of the upper part of the abdomen where it is in close association with the **alimentary canal**. The liver is an important organ and has many functions. Some of the main ones are:
(a) Production of **bile** for use in digestion.
(b) Deamination of **amino acids** which are excess to the needs of the body.
(c) Regulation of **blood sugar** by the interconversion of **glucose** and **glycogen**.
(d) Storage of **iron** and **vitamins** A and D.
(e) Detoxification of poisonous substances.
(f) Release and distribution of **heat** production by the chemical activity of liver cells.
(g) Conversion of stored **fat** for use by the tissues.
(h) Manufacture of the **plasma** protein fibrinogen which is involved in **blood clotting**.

lungs The breathing **organs** of mammals, amphibians, reptiles and birds. In mammals the lungs are two elastic sacs located in the thorax. They can be expanded or compressed by movements of the thorax in such a way that air is continually taken in (inhaled) and expelled (exhaled).

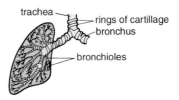

lungs *Air passages in the lung.*

   Air is taken into the body through the wind pipe (trachea). The trachea divides into two bronchi which enter the lungs. Each bronchus then further divides into smaller bronchioles, which eventually terminate as millions of air-sacs called alveoli. The alveoli are richly supplied with **blood capillaries**, and **gases** are able to diffuse into and out of the bloodstream. *See* **breathing in mammals, gas exchange**.

The smoke from cigarettes and other forms of tobacco contains substances, such as nicotine, carbon monoxide and tar, which damage the lungs and cause problems elsewhere in the body.

Nicotine is an addictive drug that causes the heart to beat faster and blood pressure to rise.

Carbon monoxide is poisonous gas that prevents the blood from carrying as much **oxygen** as it should.

Tar contains many chemicals known to cause lung cancer and other forms of cancer.

M

magma Molten material coming from the mantle area inside the **Earth** out of which **igneous rocks** crystallize. Magma is not just molten rock, it also contains *volatiles*. These are substances which evaporate easily. They include **water**, **carbon dioxide**, hydrogen sulphide, **sulphur**, sulphur dioxide and **chlorine**. Water is by far the major volatile and may account for up to 8% of the magma.

magnet A magnet is a device which strongly attracts certain **metals** (**iron**, **steel**, nickel and cobalt) and affects an electric **current**).

Magnets are often made of iron or steel but may also be made of certain **alloys** or **ceramic** materials. *Electromagnets* have similar properties to magnets and are produced by the passage of an electric current.

The areas in a magnet where the attractive **force** appears to be concentrated are called the *poles*. The two poles are called north and south. The region around the magnet, in which there is a magnetic force, is called a **magnetic field**.

The *law of magnetic force* describes the behaviour of magnetic poles: like poles repel each other while unlike poles attract each other.

A *compass* is a small magnet which is used to establish direction relative to the earth's magnetic field. The earth behaves as if there is a magnetic south pole in the region of the geographic north pole. Hence, if a magnetic needle is suspended by a thread and allowed to swing freely, its north pole will point to geographic north. For this reason it is sometimes referred to as the north-seeking pole.

magnetic field
A region of space around a **magnet** in which there is a magnetic **force**. Magnetic fields are three-dimensional but are usually represented as two-dimensional in drawings. A magnetic field is shown as a series of imaginary magnetic field lines. The

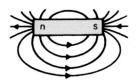
(a) Bar magnet

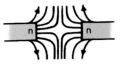
(b) Two like poles

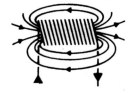
(c) A coil with current

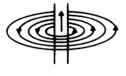
(d) A straight current

magnetic field *Sections of some simple magnetic fields.*

closeness of the field lines indicates their strength, while their direction is shown by arrow heads. Field lines are taken to run from north poles and to south poles.

mass A measure of the amount of matter in an object. It depends on the number of **molecules** it contains. The **SI unit** of mass is the kilogram, kg. The mass of an object does not vary with changes in **gravity** so an object would have the same mass of **Earth** as on the **Moon**.

   Mass should not be confused with **weight**, which changes with changes in gravity.

meiosis The formation of **gametes** or sex **cells** by the division of nuclear material in a cell **nucleus**. This process involves two consecutive divisions.

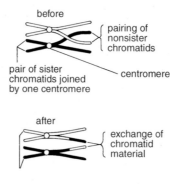

meiosis *Crossing over between non-sister chromatids.*

   Initially the cell nucleus contains the normal *diploid* number of **chromosomes**. At the start of the first division *homologous chromosomes* (chromosomes containing **genes** for the same characteristics) pair together and replication occurs. These chromosomes then separate giving two sets of the diploid number of chromosomes. During this first division homologous chromosomes may exchange genetic material as they lay side by side in a process called *crossing over*. This leads to variation in the resulting nuclei.

   In the second division each set of chromosomes divides again resulting in four *haploid* nuclei that contain only half the normal number of chromosomes. Each nucleus becomes surrounded in a portion of the cell cytoplasm forming a gamete.

interphase:
contents of nucleus
indistinct

prophase:
contents of nucleus
become clear

homologous chromosomes
form a pair.
Nuclear membrane
breaks down

each chromosome
can be seen to consist
of two chromatids
joined by a centromere

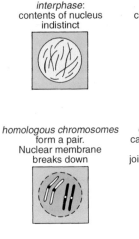

metaphase:
nuclear spindle
forms and
chromosomes
move to the equator

anaphase:
centromeres repel
each other carrying
chromosomes towards
the poles of the spindle

telophase:
nuclear membranes
form around
the groups
of chromosomes

Metaphase
Nuclear spindles form
and chromosones move
to the equator

Anaphase
Centromeres divide
and repel each other
carrying chromatias
towards the poles

Telophase
Nuclear spindles break
down and membranes
form around haploid
groups of chromosones

meiosis *Stages in the two divisions of the process.*

melting point The **temperature** at which a **solid** changes to a **liquid** (or a liquid to a solid). More precisely, it is the temperature at which solid and liquid forms of the same substance (e.g. ice/water) are in equilibrium. At constant **pressure** the melting point is constant for a pure substance, but it is lowered if impurities are added, hence ice melts when salt is sprinkled on it. The **energy** needed to cause melting without change of temperature is called **latent heat**.

menstrual cycle This is a reproductive cycle occurring in female primates (monkeys, apes and humans). The cycle is controlled by **hormones**. In human females, the cycle lasts about twenty eight days. During this time the **uterus** wall thickens in preparation for implantation of a fertilized ovum. If fertilization does not occur, the new uterus lining and the unfertilized ovum are expelled from the vagina. This part of the cycle is called *menstruation* and results in a small amount of bleeding from the vagina. *See* **fertilization** in humans, **ovulation**.

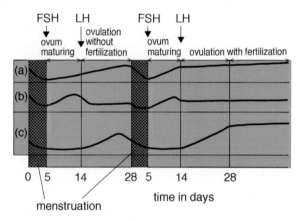

ovulation *The main features of the human menstrual cycle:*
(a) variation in thickness of wall of uterus;
(b) oestrogen level in blood (stimulates repair of the uterus wall);
(c) progesterone level in blood (prepares uterus for implantation).

mercury The only **liquid metal** at room temperatures. It has many important uses. Here are some examples:
(a) **Alloys** of mercury with other metals are called amalgams. Zinc amalgam is used for teeth fillings.
(b) Mercury is used as the liquid in **thermometers**.
(c) In mercury cells, the metal forms the cathode (negative **electrode**) in the **electrolysis** of sodium chloride solution to form sodium hydroxide.

Both mercury vapour and mercury compounds are poisonous and must be handled with extreme care.

metabolism A collective term describing all of the physical and chemical processes occurring within a living organism. These include both the synthesis (anabolism) and breakdown (catabolism) of **compounds**.

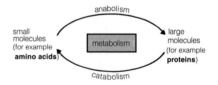

The rate of metabolism of a resting animal, measured by **oxygen** consumption, is known as the basal metabolic rate (BMR). It is the minimum amount of **energy** needed to maintain life and varies with species, age and gender.

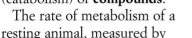

metabolism The synthesis and breakdown of compounds within an organism.

metal Many **elements** are metals. Pure metals are rarely used in modern life. Most metals are used in the form of **alloys**. Typical metals:
(a) Have a crystalline structure.
(b) Are strong and hard.
(c) Are malleable and ductile.
(d) Produce *cations* (positively charged ions).
(e) React with **acids**.
(f) React with **nonmetals**.
(g) Produce basic oxides.
(h) Good conductors of **heat** and **electricity**.
(i) Have high **melting** and **boiling points**.

metamorphic rocks Rocks which have recrystallized, in the **solid** state, from existing rocks, in conditions of **heat** and **pressure**. This process is called *metamorphism*. It does not usually change the overall chemical composition of a rock but forms new minerals from those originally present. Metamorphic rocks form from **sedimentary rocks**, **igneous rocks** or other metamorphic rocks.

Existing rock	→	Metamorphic rock
Sandstone		Quartzite
Limestone		Marble
Mudstone		Slate
Granite		Gneiss
Gneiss		Migmatite

methane A gaseous alkane with the formula CH4. In the methane **molecule** the **hydrogen atoms** are arranged in the shape of a *tetrahedron* around the **carbon** atom.

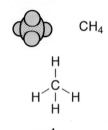

methane

Methane is the main constituent of **natural gas** and is also released from **petroleum**. It burns readily in a plentiful supply of **air** to give **carbon dioxide** and **water**, and is also an industrial source of hydrogen.

microbe A general term used to describe a microscopic organism (microorganism).

microwaves A region of the **spectrum** of **electromagnetic waves**. The approximate wavelength range is 1 mm to 10 m and the approximate **frequency** range is 10^7–10^{12} Hz. *See* **wave**.

Uses of microwaves include microwave cookers and **radar**. In microwave cookers the frequency used (typically 2450 MHz) is strongly absorbed by the food, causing it to heat up rapidly. The microwaves penetrate the food, which is cooked more quickly than in a conventional oven.

mineral salts Components of **soil** formed by the **weathering** of rocks and from **humus** mineralization. Mineral salts are found in solution in soil **water**. They are absorbed by plant roots and transported through the plant in solution. Like **vitamins**, mineral salts are needed in tiny amounts but are nevertheless vital for plant and, ultimately, animal nutrition. The absence of a particular mineral salt can lead to mineral deficiency, disease and death. Plants need at least twelve mineral salts for healthy growth.

Mineral salt	Function	Some deficiency effects
Calcium	Component of plant **cell** walls and animal **bones**.	Rickets in humans.
Iron	Component of haemoglobin.	Anaemia in humans.
Magnesium	Component of **chlorophyll**.	Pale yellow plant **leaves** (chlorosis).
Nitrogen	Component of **protein** and nucleic acids.	Poor reproductive development in plants.
Phosphorus	Components of ATP, nucleic acids, cell membrane, animal bones.	Stunted plant growth.

mineral salts

(a) *Major elements* – needed in relatively large amounts: nitrogen, phosphorus, **sulphur**, potassium, **calcium**, magnesium.

(b) *Minor elements* – needed in very small amounts: manganese, **copper**, zinc, **iron**, boron, molybdenum. Some mineral salts are needed by plants, some are needed by animals, and some are needed by both. The properties of some important mineral salts are summarized below.

mitosis The formation of new **cells** for growth and repair. Initially the **nucleus** of a cell has the normal *dipliod* number of **chromosomes**. During the process, the chromosomes replicate and then separate forming two cells. Each new cell is identical to the parent cell and contains the diploid number of chromosomes.

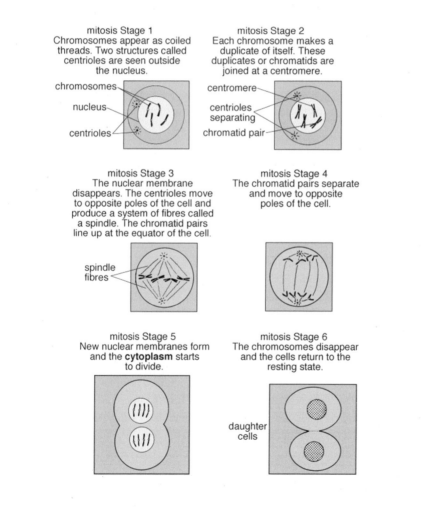

mitosis Stage 1
Chromosomes appear as coiled threads. Two structures called centrioles are seen outside the nucleus.

chromosomes
nucleus
centrioles

mitosis Stage 2
Each chromosome makes a duplicate of itself. These duplicates or chromatids are joined at a centromere.

centromere
centrioles separating
chromatid pair

mitosis Stage 3
The nuclear membrane disappears. The centrioles move to opposite poles of the cell and produce a system of fibres called a spindle. The chromatid pairs line up at the equator of the cell.

spindle fibres

mitosis Stage 4
The chromatid pairs separate and move to opposite poles of the cell.

mitosis Stage 5
New nuclear membranes form and the **cytoplasm** starts to divide.

mitosis Stage 6
The chromosomes disappear and the cells return to the resting state.

daughter cells

mixture Two or more substances present in the same container, but not chemically bonded together (*see* **compound**). Some properties of mixtures:
(a) The proportions of substances in a mixture can vary.
(b) No **energy** is absorbed or released when a mixture is made.
(c) The **properties** of a mixture are the properties of all the components.
(d) Mixtures can be separated by physical methods.

mole A mole is the **mass** of an **element** or **compound** which contains the same number of **atoms or molecules** as there are atoms in 12 g of **carbon**.

The mass of one mole of an element is the **relative atomic mass** (expressed in grams). The mass of one mole of a **compound** is the **relative molecular mass** (expressed in grams).

Symbol	H	C	O
Relative atomic masses	1	12	16
Mass of 1 mole of atoms	1 g	12 g	16 g
Formula	H_2	C	O_2
Mass of 1 mole of molecules	2 g	12 g	32 g

Thus the mass of 1 mole of water molecules is $(2 \times 1) + 16 = 18$ g, and the mass of 1 mole of carbon dioxide molecules is $12 + (2 \times 16) = 44$ g.

molecular formula The formula of a substance which shows the number and types of **atoms** found in the **molecule**. However, it does not tell us anything about the way in which those atoms are arranged (structure). For example, the molecular formula of water is H_2O, thus a molecule of water consists of two atoms of **hydrogen** and one atom of **oxygen**. It is not unusual for two substances with completely different structures and chemical properties to have the same molecular formula. For example, the chemicals butanol and ethoxyethane both have the molecular formula $C_4H_{10}O$. *See* **empirical formula, structural formula**.

molecule This is the smallest particle of an **element** or **compound** which can exist independently. It contains two or more **atoms** bonded together in small whole numbers, e.g:

O_2	a molecule of **oxygen**
CH_4	a molecule of **methane**
$CaCO_3$	a molecule of calcium carbonate

moment The turning effect, or torque, of a **force**, or couple, around a point. The moment is given by the product of the force, F, and its perpendicular distance, s, from the turning point: $T = Fs$. The **SI unit** is the newton metre, Nm.

Moments are often described as clockwise or anticlockwise, depending on the direction of the turning effect. The *law of moments* states that when an object is in equilibrium there is no net turning effect: the sum of the anticlockwise moments is equal to the sum of the clockwise moments.

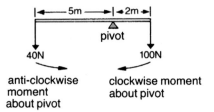

moment *In this example the anticlockwise moment and clockwise moment are both equal to 200 Nm.*

momentum The momentum of an object is given by the product of its **mass** and **velocity**. The **SI unit** is the kilogram-metre per second, kgm/s. The *law of conservation of momentum* follows from **Newton's laws of motion**. It states that the total momentum of a group of objects remains constant unless a net outside force acts on them.

monomer The **compound** from which **polymers** are made. For example:

Monomer	Polymer
Chloroethene	Polychloroethene
Ethene	Polythene
Phenylethene	Polyphenylethene
(Styrene)	(Polystyrene)

monohybrid inheritance The inheritance of one pair of characteristics for a particular feature of a plant or animal.

For example, in the fruit fly *Drosophila*, wings may be normal or vestigial (reduced in size). The **gene** for normal wings, represented by N, is *dominant* while the gene for vestigial wings, represented by n, is *recessive*. A pure-bred normal-winged fly carries the genes NN and a pure-bred vestigial fly carries the genes nn. Both are described as *homozygous* since they carry identical genes for the same feature.

If a pure-bred normal-winged fly is crossed with a pure-bred vestigial-winged fly, all of the flies in the resulting F_1 *generation* will have normal wings (their *phenotype*) and carry the genes Nn (their *genotype*). The flies are described as *heterozygous* since they carry different genes for the same feature. The trait of the dominant gene (in this case normal wings) masks the trait of the recessive gene (in this case vestigial wings).

If two flies from the F_1 generation are crossed the resulting F_2 *generation* contains a mixture of normal-winged and vestigial-winged flies in the ratio of 3:1. This ratio is determined by considering the different combinations of genes possible from the parents. In this case the **gamete** from each parent may carry either N or n. It is often convenient to show this in the form of a table known as a *Punnett square*.

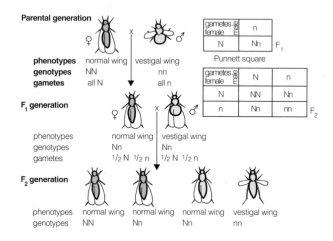

monohybrid inheritance *The results of crossing normal-winged female Drosophila with vestigial-winged males.*

moon A term used to describe a large natural satellite of any **planet**.
It is, however, most often used when describing the **Earth**'s natural satellite. The Moon is approximately 3460 km in diameter and 382 000 km from the Earth.

The Moon is a *non-luminous* source of light: it reflects light from the Sun. As the Moon revolves around the Earth differing amounts are illuminated, giving rise to the 'phases of the moon'. Half of the Moon is illuminated at all times. The inner circle shows the position of the illuminated half of the Moon. The outer

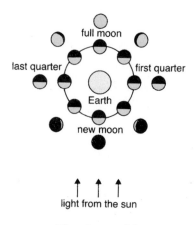

moon *The phases of the moon.*

circle shows how the Moon looks from Earth during its different phases. *See* **eclipse**.

The Moon orbits the Earth approximately once each month. *See* **tide**.

motor, electric A device which converts input electric **energy** into movement. In most electric motors movement is due to a central rotating coil. The structure of a simple motor is identical to that of a simple **generator**; however, they have opposite effects:
(a) In a generator the coil is turned and an electric **current** is produced in it.
(b) In an electric motor an electric current passes through the coil and causes it to turn.

mucus A slimy **fluid** secreted by goblet **cells** found in the epithelia of vertebrates. Mucus has several important functions in humans.
(a) It traps dust and **bacteria** in the air passages.
(b) It lubricates the surfaces of the internal **organs**.
(c) It lubricates the inside walls of the **alimentary canal**, making it possible for food to pass through easily.
(d) The layer on the inside walls of the alimentary canal prevents digestive **enzymes** from reaching and digesting the walls themselves.

multicellular (of an organism) Consisting of many **cells**. Most animals and plants are multicellular. Compare **unicellular**.

muscle Animal **tissue** which consists of **cells** which are able to contract in response to nerve impulses, thus producing movement. This may be movement of internal **organs** within the animal or movement of the animal in its **environment**.

mutation A natural spontaneous change in the structure of **DNA** in **chromosomes**. Mutations are rare, but when they occur the resulting altered characteristic is passed on to subsequent generations. Mutations most often confer disadvantages on the organism inheriting them; however, mutations may also result in beneficial **variations** within a *population*. Such beneficial variations are the basis of **evolution**.

Mutations can be induced in organisms by exposure to excessive **radiation**. The artificial alteration of DNA in chromosomes is called **genetic engineering**. Scientists are currently looking at ways of using this technique to improve the yield and immunity to disease of many food crops.

N

National Grid A network of cables, often supported in the air by pylons, which connect the country's power stations with **electricity** consumers. In general, electricity is generated in power stations at 25 kV. It is transformed (see **transformer**) to 275 or 400 kV for transmission over long distances through the National Grid. The **voltage** is reduced later by substation transformers to lower voltages suitable for domestic (240 V) or industrial use.

natural frequency The particular frequency at which any system which can vibrate will vibrate freely and naturally. The value of this depends on the physical nature of the system. For example, a simple **pendulum** will swing at a **frequency** which depends on its length and on the gravitational field strength at that place. The frequency of the note produced by blowing across the top of the bottle depends on the size and shape of the air space inside it. These examples of **resonance** are matched by others in electronics, radiation and nuclear physics.

natural gas A **mixture** containing mainly **hydrocarbon gases** which is found in deposits beneath the **Earth**'s surface. Natural gas and crude oil (see **petroleum**) are often found together. **Methane** is usually the major constituent and often makes up over 90% of the mixture. The **noble gas** helium is sometimes present in small amounts. Natural gas is used as a **fuel** in both industry and the home.

natural selection A theory first proposed by the famous scientist Charles Darwin to suggest how **evolution** could have taken place. Darwin suggested that individual organisms within a **species** differ in the extent to which they are adapted to conditions of their **environment**. Thus, in competition for food etc., the better adapted organisms will survive and pass on their favourable **variations** to future generations. Conversely, the less well-adapted will die out.

nervous system A network of specialized **cells** in **multicellular** animals. This network links receptors and effectors, thus coordinating the animal's activity. In mammals, the nervous system consists of the **brain** and the **spinal cord** (which together are called the central nervous system) and nerve cells connecting to all parts of the body.

neutralization A process where either the **pH** of an acid solution is increased to 7 or the pH of an alkaline solution is decreased to 7. pH 7 is neutral. An acid can be neutralized by adding a base or a compound such as a *carbonate*:

$$HCl(aq) + NaOH(aq) \rightarrow NaCl(aq) + H_2O(l)$$
$$2HCl(aq) + CuCO_3(s) \rightarrow CuCl_2(aq) + CO_2(g) + H_2O(l)$$

In acid-base reactions, **indicators** are used to show when the neutralization has occurred.

neutron One of the three main *subatomic particles* found in most **atoms**. The neutron has the same **mass** as a **proton** but carries no electrical **charge**.

Newton's laws of motion Three important statements which relate **forces** to each other and their effects on objects. They are valid for all forces and all cases, and form the basis of much physics.

First law: A body stays at rest, or if moving it continues to move with uniform **velocity**, unless an external force acts on it making it behave differently.

Second law: The rate of change of **momentum** equals the net force. This is often written simply as:

$$\text{Force} = \textbf{mass} \times \textbf{acceleration} \ (F = m \times a).$$

Third law: When object A applies a force on object B, then B applies an equal but opposite force on A. This is often expressed as 'action and reaction are equal and opposite'.

The second law helps to define the newton, N, the unit of force. It is the force which will accelerate a mass of 1 kg by 1 m/s^2.

nitrogen cycle The circulation of the **element** *nitrogen* and its **compounds**. Much of this cycle is the result of metabolic processes (see **metabolism**) of living organisms. However, there are large amounts of synthetic nitrogen-containing **fertilizers** made from atmospheric nitrogen (via **ammonia** and nitric acid), used each year. These also play a significant part. Nitrates may also result from the action of lightning on atmospheric gases. See diagram overleaf.

noble gases or inert gases **Elements** in **Group** 0 of the **Periodic Table**. They are all monatomic **gases** found in small amounts in the atmosphere, i.e. helium, neon, argon, krypton, xenon.

Their electronic configurations are very stable because they have a

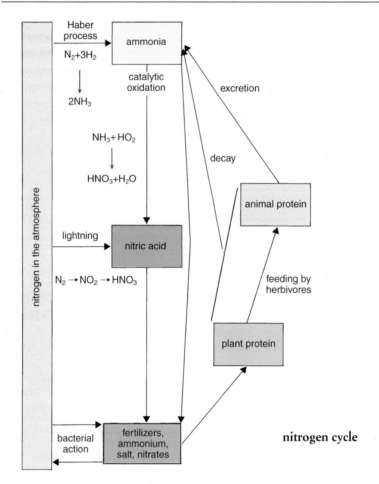

nitrogen cycle

complete outer shell of **electrons**. The elements do not easily lose or gain electrons. A few chemical **compounds** exist which contain **atoms** of the more massive noble gases (e.g. XeF_4) but these are man-made. When elements form **ions**, the electrons of these ions have an inert gas configuration, e.g:

Element	Ion	Electronic configuration atom	ion	Noble gas equivalent
Sodium	Na^+	2·8·1	2·8	Neon
Fluorine	$F-$	2·7	2·8	Neon
Calcium	Ca^{2+}	2·8·8·2	2·8·8	Argon
Sulphur	S^{2-}	2·8·6	2·8·8	Argon

nonmetals **Elements** which either:
(a) Have molecular structures and are **gases** at room temperature, or are **solids** or **liquids** with low **melting** and **boiling points**.
(b) Have **giant structures** with **covalent bonding**.
Typical nonmetals:
(a) Are poor conductors of heat and electricity.
(b) Give rise to **anions** (negatively charged ions).
(c) Do not react with **acids**.
(d) Produce acidic oxides.
(e) Form **covalent compounds**.

nuclear power The supply of useful **energy** from nuclear **fission** reactions in a *nuclear reactor*. The energy is produced by a controlled **chain reaction**. The **kinetic energy** of the particles produced by each fission produces a rise in **temperature**. A coolant removes the **heat** energy to make steam. The steam drives turbines which, in turn, drive **generators**.

nucleus **1.** The part within most **cells** which contains **chromosomes**. As the chromosomes contain the hereditary information, the nucleus controls the cell's activity through the action of the genetic material, **DNA**. The nuclear membrane isolates the nucleus from *cytoplasm*.

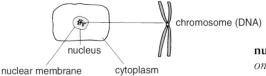

chromosome (DNA)

nucleus

nuclear membrane   cytoplasm

nucleus *The nucleus of a cell and one of its chromosomes.*

2. The part at the centre of an **atom** where the **mass** is concentrated. It contains **protons** and **neutrons**. The nucleus has a positive **charge**, and in the neutral atom this is balanced by negatively charged **electrons** moving

proton —   — neutron

nucleus *Nucleus of a lithium atom.*

around the nucleus. Atoms of the **isotope** hydrogen-1 are the only atoms whose nuclei contain no neutrons.

nylon The nylons are a group of *polyamide polymers*. The commonest of them is called nylon 6.6. *See* **polymerization**.
  This synthetic *fibre* has great strength. It is used in fabrics, yarns (stockings, knitwear), carpets, ropes and nets. Nylon is useful because it

will not rot, is hard-wearing and does not absorb water; however, it does stretch (useful in ropes and stockings).

Nylon is often mixed with other fibres such as wool to get a balance of properties. Generally speaking a nylon jumper is not as warm as a woollen one, however it will not wear out as quickly. A mixture of fibres produces a jumper which is both warm and hard-wearing.

O

oestrogen A **hormone** which is secreted by the **ovaries** of vertebrates. It stimulates the development of **secondary sexual characteristics** in female mammals, and plays an important role in the **menstrual cycle**.

oil A **liquid fat** such as melted butter, olive oil, sunflower oil, etc. The word is sometimes used in place of **petroleum**.

oleum A solution of sulphur(VI) oxide (sulphur trioxide) in concentrated **sulphuric acid**. Oleum is extremely corrosive and is a vigorous oxidizing agent.

optic nerve A cranial nerve found in vertebrates. It conducts nerve impulses from the retina of the eye to the **brain**.

orbit The closed path of an object moving in a central force field. For example, the **Moon** travels in an orbit around the **Earth** moving in the Earth's gravitational field. The central force field is continually pulling the object towards it, thus preventing it from moving in a straight line. The object has both **potential** and **kinetic energy**.

ore A naturally occurring substance from which an **element** (often a **metal**) can be extracted.

Element	Ore
Aluminium	Bauxite
Chlorine	Rock salt
Copper	Chalcopyrite ($CuFeS_2$)
Iron	Haematite (Fe_2O_3)
Lead	Galena (PbS)
Zinc	Zinc blende (ZnS)

ore *Elements and the ores from which they are extracted.*

organ A collection of different **tissues** in a plant or animal which forms a structural and functional unit within an organism, such as the **leaf** of a plant or the **liver** of an animal. Different organs may be associated together to constitute a system; for example, the digestive system.

$$\text{cells} \rightarrow \text{tissues} \rightarrow \text{organs} \rightarrow \text{systems}$$

organic chemistry This branch of chemistry is concerned with the study of **compounds** of **carbon** but does not include carbonates, hydrogencarbonates, **carbon dioxide**, etc.

osmosis The **diffusion** of a solvent (usually water) through a selectively permeable membrane from a region of high solvent **concentration** to a region of lower solvent concentration. Examples of selectively permeable membranes are:
(a) The **cell** membrane.
(b) Visking (dialysis) tubing.
Selectively permeable membranes are thought to have tiny pores which allow the rapid passage of small **water** molecules but restrict the passage of larger solute molecules.
Since the cell membrane is selectively permeable, osmosis is important in the passage of water into and out of cells and organisms. The **pressure** exerted by the movement of water due to osmosis is called *osmotic pressure*.

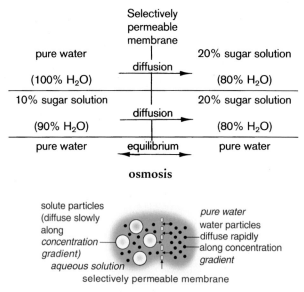

osmosis *Passage of solvent (water) particles through a selectively permeable membrane.*

ovary 1. The hollow region in the **carpel** of a **flower** which contains one or more ovules. *See* **fertilization** in plants.
2. The reproductive **organ** of female animals. In vertebrates there are two ovaries. They produce **ova** and also release certain sex **hormones**.

ovulation A process which involves the release of an **ovum** onto the surface of a vertebrate **ovary**. From here it passes into the oviduct and then into the **uterus**.

In the human female, ovulation is controlled by **hormones** secreted by the pituitary gland. The sequence of events in the female's reproductive behaviour is called the **menstrual cycle**.

ovum An unfertilized female **gamete** produced at the **ovary** of many animals. An ovum contains a *haploid* **nucleus**.

oxidation A substance is oxidized if it:
(a) Gains **oxygen**, e.g:

$$2Mg(s) \quad + \quad O_2(g) \quad \rightarrow \quad 2MgO(s)$$
magnesium   oxygen   magnesium oxide

(b) Loses **hydrogen**, e.g:

$$CH_4(g) + Cl_2(g) \rightarrow CH_3Cl(l) \quad + \quad HCl(g)$$
methane   chlorine   chloromethane   hydrogen
chloride

(c) Loses **electrons**, e.g:

$$Cu(s) \quad \rightarrow \quad Cu^{2+}(aq) \quad + \quad 2e$$
copper   copper ions   electrons

A substance which brings about the oxidation of another substance is called an oxidizing agent. Compare **reduction**.

oxygen (O_2) A gaseous nonmetallic **element** in **Group** VI of the **Periodic Table**. The **gas** is both colourless and odourless and makes up 21% of the **atmosphere**. It is vital for the **respiration** of both plants and animals. Although atmospheric oxygen is used up during respiration the supply is continually replenished by plants as a result of **photosynthesis**. Oxygen is only slightly soluble in **water**; however, there is normally enough dissolved to support fish and other aquatic life. **Pollution** often causes a drop in the **concentration** of oxygen in the water to such an extent that fish will suffocate and die. Oxygen is obtained industrially by the *fractional distillation* of liquefied air. In the laboratory it is obtained by the decomposition of hydrogen peroxide, usually with the help of a **catalyst** such as manganese(IV) oxide:

$$2H_2O_2(aq) \rightarrow 2H_2O(l) + O_2(g)$$

The chemical test for oxygen is to put a glowing splint into the gas. If the gas is oxygen the splint will relight. Oxygen will also make anything already burning in air burn much more fiercely. Oxygen is a vigorous *oxidizing agent*. It is used extensively in **steel**-making and welding, as a rocket propellant together with kerosine or **hydrogen**, and for life-support systems in medicine.

ozone layer A region of the **atmosphere** about 30 km above the **Earth's** surface, which is rich in ozone. (The **gas** ozone is an *allotrope* of **oxygen** where the **molecule** is triatomic (O_3).) This layer prevents most **ultraviolet radiation** from reaching the Earth. This is important, as exposure to excess ultraviolet radiation is known to cause skin cancer.

Scientists have recently become aware of holes appearing in the ozone layer allowing unusually large amounts of ultraviolet radiation to reach the earth. Chemicals called *chlorofluorocarbons* (CFCs), used as aerosol propellants and in many industrial processes, are thought to be the cause. Some CFCs eventually diffuse up into the ozone layer. They are very unreactive chemicals under normal room conditions; however, in the presence of sunlight they react with the ozone, thus destroying it. Steps are now being taken to reduce the amounts of CFCs used in the world.

P

pacemaker A region of the *vertebrate* **heart** where the contraction at each **heartbeat** starts. This word is often used to refer to an electronic device which may be inserted into the chest of a person suffering from certain types of heart disease. The device provides the heart with the electrical impulses needed for a regular beat.

pancreas A gland situated near the duodenum (see **alimentary canal**) of vertebrates. It releases an alkaline fluid containing digestive **enzymes** such as lipase, amylase and trypsin into the duodenum (see **digestion**). The pancreas also contains **tissue** known as the *Islets of Langerhans* which secretes the **hormone insulin**, which is important in the **metabolism** of **sucrose**.

parasite An organism that feeds in or on another living organism called the *host*. In a parasitic relationship the host does not benefit and may actually be harmed. Examples in man include fleas, lice and tapeworms.

particle (In physics) an object whose **volume** is very small. The **kinetic model** describes the structure of matter in terms of particles. In simple pictures of matter, such particles are called **atoms** or **molecules**. Studies have revealed that particles can sometimes appear as **waves** because they can show wave properties like **diffraction** and interference.

pascal (Pa) The **SI unit** of **pressure**. 1 pascal = 1 newton/metre^2.

pendulum A regularly swinging object with a regular transfer to and fro between **potential** and **kinetic energy**.

The period of a pendulum is the time it takes to complete one full cycle.

The value of the period, T, depends only on the length, *l*, of the pendulum and the acceleration due to gravity, g. The value of g can be estimated by carrying out a simple experiment with a pendulum and putting the results into the following equation:

$$g = \frac{4 \times \pi^2 \times l}{T^2}$$

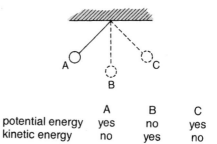

	A	B	C
potential energy	yes	no	yes
kinetic energy	no	yes	no

pendulum *The period of this pendulum is the time it takes to swing from A to C and back to A*

periodic table A table representing all of the **elements** so as to show similarities and differences in their chemical properties.

The elements are arranged in increasing order of **atomic number** (Z) as you read from left to right across the table. The horizontal rows are called periods and the vertical columns are called **groups**. There is a progression from **metals** to **nonmetals** across the period. This is Period 3:

element	Na	Mg	Al	Si	P	S	Cl	Ar
atomic number	11	12	13	14	15	16	17	18
		metals					nonmetals	

This is part of Group I:

Li
Na
K

The block of elements between Groups II and III are called *transition metals*. The chemistry of these metals is similar in many ways. Elements 58–71 are known as the *Lanthanide* or *Rare Earth* elements. Elements 90–103 are known as the *Actinide* elements. The elements which have atomic numbers greater than 92 do not occur naturally but have been made artificially by bombarding other elements with *particles*.

pest A general term used to describe any organism which is considered to have a detrimental effect on humans. This effect may be to the body, to the food supply or to the **environment**. Some examples are shown in the following table:

Example of pest	Effect on humans
Weeds, locusts	Reduce the growth of plants and crops
Foot and mouth virus	Causes disease in domestic animals
Woodworm, wet rot fungus	Damages buildings
Mosquitoes, lice	Transmit human disease

Methods used to combat pests include:
(a) Spraying with chemical **pesticides**.
(b) Using natural **predators** against the pest.
(c) Introducing **parasites** and disease-bearing organisms to the pest population.
(d) Introducing sterile (unable to breed) individuals to the pest population, thus reducing the reproductive capacity.

periodic table

groups	1	2												3	4	5	6	7	0(8)
periods																			
1							1 H hydrogen 1.0												2 He helium 4.0
2	3 Li lithium 6.9	4 Be beryllium 9.0												5 B boron 10.8	6 C carbon 12.0	7 N nitrogen 14.0	8 O oxygen 16.0	9 F fluorine 19.0	10 Ne neon 20.2
3	11 Na sodium 23.0	12 Mg magnesium 24.3												13 Al aluminium 26.9	14 Si silicon 28.1	15 P phosphorus 31.0	16 S sulphur 32.1	17 Cl chlorine 35.5	18 Ar argon 39.9
4	19 K potassium 39.1	20 Ca calcium 40.1	21 Sc scandium 45.0	22 Ti titanium 47.8	23 V vanadium 50.9	24 Cr chromium 52.0	25 Mn manganese 54.9	26 Fe iron 55.9	27 Co cobalt 58.9	28 Ni nickel 58.7	29 Cu copper 63.5	30 Zn zinc 65.4		31 Ga gallium 69.7	32 Ge germanium 72.6	33 As arsenic 74.9	34 Se selenium 79.0	35 Br bromine 79.9	36 Kr krypton 83.8
5	37 Rb rubidium 85.5	38 Sr strontium 87.6	39 Y yttrium 88.9	40 Zr zirconium 91.2	41 Nb niobium 92.9	42 Mo molybdenum 95.9	43 Tc technetium 98	44 Ru ruthenium 101.1	45 Rh rhodium 102.9	46 Pd palladium 106.4	47 Ag silver 107.9	48 Cd cadmium 112.4		49 In indium 114.8	50 Sn tin 118.7	51 Sb antimony 121.8	52 Te tellurium 127.6	53 I iodine 126.9	54 Xe xenon 131.3
6	55 Cs caesium 132.9	56 Ba barium 137.3	57 La lanthanum 138.9	72 Hf hafnium 178.5	73 Ta tantalum 181.0	74 W tungsten 183.9	75 Re rhenium 186.2	76 Os osmium 190.2	77 Ir iridium 192.2	78 Pt platinum 195.1	79 Au gold 197.0	80 Hg mercury 200.6		81 Tl thallium 204.4	82 Pb lead 207.2	83 Bi bismuth 209.0	84 Po polonium 209	85 At astatine 210	86 Rn radon 222

transition metals

key

atomic no. Symbol
name
relative atomic mass

metal

non metal

transition metal

metalloid

pesticide A chemical **compound**, often delivered as a spray or fine powder, which kills or prevents the **growth** of pests which damage crops. Pesticides are often divided into three groups:
(a) *Herbicides*. Chemicals which act on plants (weedkillers), e.g. paraquat.
(b) *Fungicides*. Chemicals which act on fungal growths, e.g. Cheshunt compound.
(c) *Insecticides*. Chemicals which act on insects, e.g. DDT (now banned in Britain).

Pesticides are used to increase the yield of crops, however there are certain disadvantages associated with their use:
(a) They may kill organisms other than the target pest.
(b) The **concentration** of pesticide increases as it passes through a **food chain**.
(c) Some pesticides decompose very slowly and may accumulate into harmful doses within consumer organisms.
(d) By killing off susceptible organisms, they allow resistant individuals to grow and multiply with reduced competition.

petroleum The **mixture** of **hydrocarbons**, e.g. **natural gas** and *crude oil*, found in the Earth's crust. It is thought to have been produced over millions of years by the action of **heat** and **pressure** on the remains of marine animals and plant organisms.

There are large reserves of petroleum in the Middle East, United States, Soviet Union, Central America and North Sea. Petroleum is the raw material of the petrochemical industry and is the source of petrol, diesel fuel, heating oil, fuel oil and gas supplies. These products are obtained by the *fractional distillation* of petroleum. This method of separation is possible because the different components, or fractions, of the mixture boil at different **temperatures**.

In recent years petroleum has replaced **coal** as the major source of raw materials for the chemical industry. However, scientists are aware that the supply of petroleum cannot last forever and the search is now on to find alternative sources.

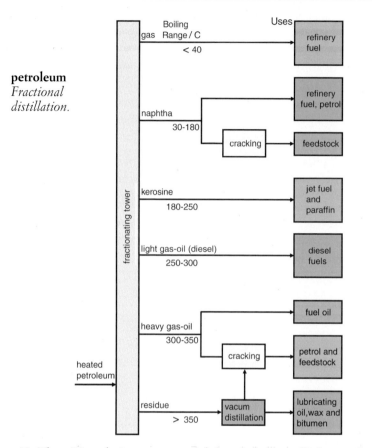

petroleum
Fractional distillation.

pH The pH scale is a measure of the *acidity* or *alkalinity* of a solution. The lower the value, the more acidic the solution, i.e. the higher the **concentration** of **hydrogen ions** (H^+) it has. A neutral solution has equal concentrations of hydrogen and **hydroxide ions** (OH^-) and has a pH of 7.

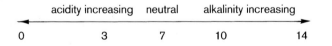

phenotype *See* **monohybrid inheritance**

photosynthesis The process by which green plants make **carbohydrate** from **carbon dioxide** and **water**. The **energy** for the reaction comes from sunlight which is absorbed by the green pigment **chlorophyll** found in the **chloroplasts**. **Oxygen** is produced as a byproduct in this process.

$$carbon+water \atop dioxide \atop 6CO_2+6H_2O \quad \xrightarrow[\text{chlorophyll}]{\text{light energy}} \quad {carbohydrate+oxygen \atop C_6H_{12}O_6 \quad + \quad 6O_2}$$

Photosynthesis is the source of all food and the basis of **food chains**, while the release of oxygen replenishes the oxygen content of the **atmosphere**. The process is thus essential to the functioning of the **biosphere**.

physical change A change to a substance which involves changes in its physical properties, e.g. **temperature**, particle size, state, etc., but no alteration to its chemical properties. Compare **chemical change**.

placenta An **organ** which develops in the uterus of a female mammal during **pregnancy**. It allows a close association between the blood circulation of the mother and that of the foetus. The placenta allows the passage of food and oxygen from the mother to the foetus and the removal of the waste products **carbon dioxide** and **urea** from the **fetus**.

planet One of nine known members of the solar system which revolve around the Sun. *See* Appendix F.

plankton The collective word used for microscopic animals and plants which float in the surface waters of lakes and seas. Plankton play a vital role as the starting point of aquatic **food chains**.

plasma The clear fluid of vertebrate blood in which the blood cells are suspended. It is an aqueous solution containing many dissolved **compounds** which are transported to different parts of the body, e.g:
(a) Waste products: **carbon dioxide, urea.**
(b) Digested foods: **glucose, amino acids.**
(c) **hormones**, plasma **proteins**, sodium chloride.

plastic 1. The nature of an object when it has been deformed beyond its *elastic limit*. Compare **elasticity**.
2. A substance which can be formed into a desired shape very easily. *See* **thermoplastic, thermosetting plastic**.

polarization Restricting the vibrations during transverse wave motion to a single plane. During transverse wave motion, the *vibration* is perpendicular to the direction in which the wave is travelling. There are many different directions in which this can occur.
  Like all **electromagnetic waves**, **light** can be polarized. Some sunglasses polarize light passing through them, reducing glare. Some

minerals are characteristic colours under polarized light and this is used to identify them.

pollen The reproductive spores of flowering plants. Each pollen grain contains a male **gamete**. Pollen grains may be transferred either by insects or by the wind, and they are adapted to their particular mode of transfer.

(a)

air bladders

(b)

pollen *(a) Spiky and sticky for insect pollination. (b) Smooth and light for wind pollination.*

pollination The transfer of **pollen** grains from the **stamens** (male part of a **flower**) to **carpels** (female part of a flower) in flowering plants.
(a) *Self-pollination* involves the transfer of pollen within the same flower or between flowers on the same plant.
(b) *Cross-pollination* involves the transfer of pollen between two separate plants.
   Normally the male and female parts of a plant do not mature at the same time, so cross-pollination is more likely. This results in the mixing of **chromosomes** and may lead to **variation**. Pollen is transferred either by the wind or on the bodies of insects. Flowers are adapted to favour one particular method of pollen transfer.
(a) *Insect pollination* – the flowers are generally highly coloured and/or scented and contain nectar. Insects are attracted by the colour and scent and visit the flower to collect the nectar. Their bodies become dusted with pollen. Some of this pollen may stick to the stigmas of subsequent flowers which they visit.
(b) *Wind pollination* – the flowers are often drab and have no scent. They produce many more pollen grains than insect-pollinated plants, as much of the pollen is lost in transfer.

pollution A general word for the addition of any substance to an **environment** which upsets the natural balance. Pollution has resulted mainly from *industrialization*. Two important factors have been a large increase in the **combustion** of **fossil fuels** and a migration of people from the land to towns and cities. Pollution of the **air** and **water** are often singled out as particularly important (see **biosphere**).
(a) Air pollution is caused mainly by the combustion of fossil fuels.
   Air pollutants such as smoke and sulphur dioxide cause irritation to the

human respiratory system and may accelerate diseases such as bronchitis and lung cancer. Sulphur dioxide and nitrogen oxides are responsible for **acid rain**. Lead is thought to retard the mental development of children.

coal burning $\longrightarrow$ smoke + carbon dioxide + sulphur dioxide

petrol burning $\longrightarrow$ smoke + carbon monoxide + oxides of nitrogen

+ lead

(b) Water pollution is caused by the intentional or accidental addition of materials into both freshwater and seawater. Most pollutants originate from industrial and agricultural practices and from the home. For example, mine and quarry washings, **acids**, **pesticides**, **fertilizers**, oil, radioactive discharges, detergents, hot water (from power stations). Some pollutants, such as pesticides, may poison and kill aquatic organisms. Other pollutants, such as sewage, cause an increase in the population of microorganisms in the water. The result is a reduction in the dissolved oxygen level, making it impossible for other organisms to survive. This process is called *eutrophication*.

polymers Large **molecules** in which a group of **atoms** is repeated, for example:

$$x–x–x–x–x–x–x–x$$
or
$$x–y–x–y–x–y–x–y$$

Polymers may be naturally occurring substances such as **starch** and **cellulose**, or synthetic substances such as **nylon** and *polythene*. See **polymerization**.

polymerization Any process in which small **molecules** (**monomers**) combine together to form large molecules (**polymers**).

Addition polymerization involves one kind of molecule containing a carbon-carbon double bond. Electrons are transferred from the double bonds allowing the molecules to join together to form a long chain of carbon-carbon single bonds. See diagram on page 206.

Condensation polymerization involves two kinds of molecule which condense together to form long chains. A small molecule is lost during the reaction. This is often, but not always, **water**. See diagram on page 117.

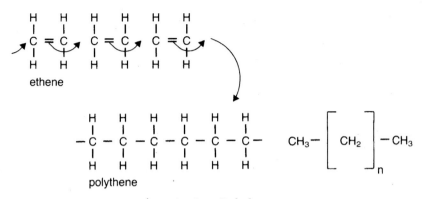

ethene

polythene

polymerization *Polythene.*

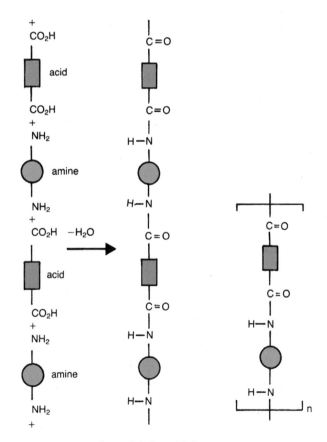

polymerization *Nylon.*

potential energy The **energy** which an object has stored up because of its position. For example, **water** at the top of a waterfall has potential energy. As the water drops, the potential energy changes into **kinetic energy**. Potential energy can also be stored as elastic, chemical or electrical energy.

potential difference *See* **voltage**.

power (P) The rate of **energy** transfer or the rate of working. The unit, the **watt**, is the transfer of one **joule** per second. The horsepower was once used as a unit of power; one horsepower is equivalent to just under 750 watts. Here are some useful equations involving power:

$$\text{mechanical power} = \textbf{force} \times \textbf{velocity} = \frac{\text{work done}}{\text{time taken}} = Fv$$

$$\text{electrical power} = \textbf{voltage} \times \textbf{current} = VI$$

Efficiency is often expressed in terms of power:

$$\text{efficiency} = \frac{\text{useful output power}}{\text{total input power}}$$

In the case of a **transformer**:

$$\text{efficiency} = \frac{P_2}{P_1} = \frac{V_2 \times I_2}{V_1 \times I_1}$$

precipitate (ppt) The insoluble substance formed when two solutions are mixed in a double decomposition reaction. In each of the following examples the precipitate is <u>underlined</u> and appears as a **solid** when the two clear solutions are mixed.

$Pb(NO_3)_2(aq)$	+	$2NaCl(aq)$	$\rightarrow$	$PbCl_2(s)$	+	$2NaNO_3(aq)$
lead(II)		sodium		<u>lead(II)</u>		sodium
nitrate		chloride		<u>chloride</u>		nitrate
$BaCl_2(aq)$	+	$Na_2SO_4(aq)$	$\rightarrow$	$BaSO_4(s)$	+	$2NaCl(aq)$
barium		sodium		<u>barium</u>		sodium
chloride		sulphate		<u>sulphate</u>		chloride

predator An animal which feeds on other animals which are called the *prey*. A predator is a food consumer but not a **parasite**. The relationship between predator and prey is reflected in their numbers. The numbers of predators and prey in an area may follow a cycle.

We can interpret the cycle as follows:
(a) When prey is plentiful a larger number of predators will survive as there is less competition for food.
(b) More predators will eat more prey, so the number of prey will decline.

(c) Fewer prey enable fewer predators to survive as there is now more competition for food.

(d) Fewer predators will eat fewer prey, so the number of prey will increase.

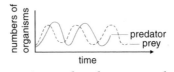

predator *A typical predator–prey relationship.*

pregnancy or gestation period The time between conception and birth in mammals. Human pregnancy lasts for approximately forty weeks and during this time the **embryo**, which has become implanted in the **uterus**, develops. In the early stages of pregnancy, finger-like structures (called *villi*) grow from the embryo into the uterus wall and develop into the **placenta**. This allows the transfer of **oxygen** and food from the mother to the embryo and the removal of **carbon dioxide** and **urea**. During the pregnancy the **cells** of the embryo continually divide and differentiate. The growing embryo (now called a **fetus**) becomes suspended in a water-filled sac called the *amnion*. The placenta extends into the umbilical cord which connects with the abdomen of the fetus.

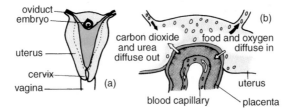

pregnancy *(a) Implanted embryo.*
(b) Relationship between uterus and placenta.

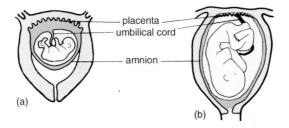

pregnancy *(a) Human fetus after twenty weeks.*
(b) Human fetus just before birth.

pressure (p) The **force** on a unit area of a surface from the substance in contact with it. The unit of pressure is the **pascal** (Pa) (1 Pa =1 N/m²). To calculate pressure we use the equation:

$$\text{pressure (Pa)} = \text{force (N)} / area \text{ (m}^2).$$

Often the force applied to an object is **weight** W. The pressure at a depth in a **liquid** or in the **air** equals the weight above the unit area. This gives the relationship:

$$\text{pressure} = \text{depth} \times \text{mean } \textbf{density} \times \text{g}.$$

The pressure exerted by the air is called *atmospheric pressure*. Pressure may be measured by a variety of instruments, e.g. manometer, Bourdon gauge, barometer.

primary sexual characteristics The features which distinguish between males and females from the time of their birth. Primary sexual characteristics do not include those which develop at puberty and are characteristic of adulthood. Compare **secondary sexual characteristics**.

progesterone A **hormone** secreted by the **ovaries** of a mammal. It prepares the **uterus** to receive the fertilized embryo and prevents any further **ovulation** during the **pregnancy**.

properties The characteristic ways in which a substance behaves (reacts) that make it what it is and make it different from other substances. Properties are often classified as *physical* or *chemical*. Chemical properties relate to the chemical reactions of the substance.

Physical properties	Chemical properties
Colour	Whether the substance is a metal
Density	or non-metal
Physical state	Gives acidic or basic oxides
Boiling point	Has more than one valency
Melting point	Reacts with acids
Crystal form	Is an oxidizing or reducing agent
Solubility	
Hardness	

properties *Physical and chemical properties*

proteins Organic **compounds** containing the **elements carbon, hydrogen, oxygen**, nitrogen and sometimes **sulphur**. A simple **molecule** of protein consists of a long chain of subunits called **amino acids**. These chains may be joined to other chains and folded in several different ways, resulting in very large and complex molecules.

Proteins are the building blocks of **cells** and **tissues**, being important constituents of **muscle, skin, bone**, etc. Proteins also play a vital role as **enzymes**, and some **hormones** also have a protein structure.

proton A positively charged subatomic **particle** found in the **nucleus** of the **atom**. The number of protons in an atom is given by the **atomic number**. **Isotopes** of any **element** always contain the same number of protons.

pulley A type of machine which transfers an applied force by means of the *tension* in a rope, cable or chain. One or more pulley wheels change the direction and/or the size of the applied force. Pulleys are often assumed to be 100% efficient (see **efficiency**) for the purpose of calculations. However, some **energy** is 'wasted' due to **friction** in the pulley bearings and stretching of the rope or cable.

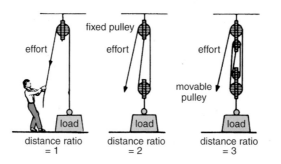

pulley *The addition of pulley wheels increases the force ratio.*

pulse rate The regular beating of blood in **arteries** due to the **heartbeat**. Pulse rate can be detected in the human body where an artery runs close to the **skin**. The wrist is often used for taking someone's pulse. In an adult human, pulse rate varies between about 70 beats per minute at rest to over 100 beats per minute during exercise.

pyramid of numbers A diagram showing the proportion of organisms at each stage in a **food chain**.

As **energy** passes along a food chain some is lost at each link because each organism uses energy in various activities such as movement. Consequently, the remaining energy can only support a progressively smaller number of organisms. Energy losses are greater when mammals are involved in a food chain because some energy from **respiration** is used to generate heat in order to maintain body temperature.

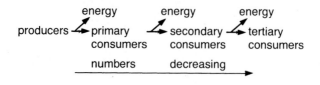

pyramid of numbers *Energy loss.*

The energy loss may also be shown as a *pyramid of biomass* in which each level corresponds to the total **mass** of organisms, or *biomass*, at that level.

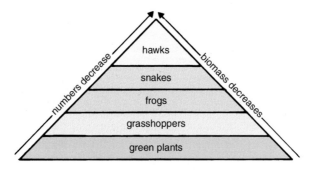

pyramid of numbers

R

radar A technique for finding the direction and distance of an object by analysing its **reflection** of **microwaves**.

Pulses of waves are emitted in different directions. Waves which hit objects are reflected back to a detector. The direction from which the waves are reflected gives the direction of the object. From knowing the speed of the microwaves, and the time taken to reach the object and be reflected back to the detector, it is possible to calculate the distance to the object.

If the object is moving there is a **frequency** change during reflection; this is called the *Doppler effect*. The size of the frequency change gives the object's speed. This method is used by traffic police to measure the speed of vehicles.

radiation The transfer of **energy** by either **waves** or particles. Any radiating object therefore loses energy. **Heat** radiation is energy transfer by **infrared** waves. It leaves all objects at a rate that depends on **temperature**, however, black surfaces radiate (and absorb) these waves better than white ones unless the temperature is very high. Cosmic radiation consists of a stream of very fast particles. The source of cosmic radiation is not known for certain, but is thought to be exploding **stars**. *See* **electromagnetic waves**, **radioactivity**.

radio A region of the **spectrum** of **electromagnetic waves**. The approximate wavelength range is 10–10^6 m and the approximate **frequency** range is 10^2–10^7 Hz. The most important use of radio is in telecommunications.
The following table lists the main bands used:

Band		Wavelength range (m)
LF	low frequency	10^4–10^3
MF	medium frequency	10^3–10^2
HF	high frequency	10^2–10
VHF	very high frequency	10–1
UHF	ultra high frequency	1–0.1

Part of the VHF and UHF bands are in the **microwave** region of the electromagnetic spectrum; however, they are often included with radio as they are also used for telecommunication.

Radio communication is helped by the *ionosphere*. This is a region of the upper **atmosphere**, from about 40–400 km above the ground. Many of the **air particles** in the ionosphere are ionized (hence the name), mainly by **ultraviolet radiation** from the **Sun**. The ionized layer reflects the radio waves.

Radio waves which have wavelengths of less than 10m are used to carry television programmes. Unfortunately these are not reflected by the ionosphere, and communication of live television programmes across the world depends on reflection from suitably placed satellites.

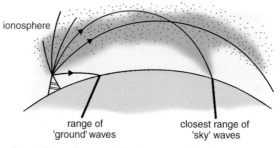

radio *The reflection of radio waves by the ionosphere.*

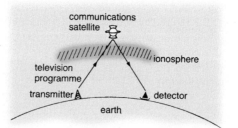

radio *Television programmes have to be reflected by satellite.*

radioactivity The decay (breakdown) of unstable **nuclei** into more stable forms. During the **decay energy** is transferred to the products; they move away; they have **kinetic energy**. There are several forms of radioactivity:

(a) Alpha decay involves the release of **alpha particles**. The alpha particle, α, is a helium nucleus and has a **mass** of four units. When a parent nucleus, X, changes to a daughter nucleus, Y, the mass of Y is four units less than that of X and its **proton** number is two less.

(b) There are several forms of beta decay. In the most common, a **neutron** in the parent nucleus changes to a proton and an **electron beta particle**, β, escapes.

(c) **Gamma radiation**, γ, is the result of a simple change of energy structure in the nucleus.

(d) In the case of **fission**, the parent nucleus splits into two smaller daughters with the release of a few neutrons. A **chain reaction** may occur.

(e) A few radioactive nuclei become more stable by the release of protons.

In all cases the daughter nuclei may themselves be radioactive, and may produce a different form of radioactivity from the parent nucleus. For this reason it is difficult to provide a pure source of any one form of radioactive **radiation**. *See* **half-life**.

rate of reaction The rate (or **speed**) at which a chemical reaction proceeds. This depends on several factors:

(a) **Temperature**. The higher the temperature, the faster the reaction, because the **particles** are moving with greater **energies**.

(b) *Particle size.* The smaller the particles involved, the greater the total surface area available for reaction, hence the faster the reaction.

(c) **Concentration**. The more concentrated a solution (or the higher the **pressure** of a **gas**) the more particles there are in a given **volume**. The more particles are available for reaction, the faster the reaction will be.

(d) **Catalysts**. These change the rate by providing an alternative reaction pathway along which the reaction can A series of elements arranged proceed.

reactivity series A series of **elements** arranged in order of their chemical reactivity.

For example, the following series shows a group of metals in order of decreasing activity:

Potassium
Sodium
Calcium
Magnesium
Aluminium
Zinc
Iron
Lead
Copper
Silver

The position in the series indicates the reactivity of an element in chemical reactions. For example, potassium reacts violently with **water**, magnesium reacts very slowly with water but more vigorously with steam. Copper reacts with neither water nor steam. *See* **electrochemical series**.

rectifier A device which allows **current** in only one direction. Most forms are types of **diodes**. Rectifiers are used to convert **alternating current** into **direct current**.

A smoothing capacitor makes the signal more steady, however the output may still not be steady enough for all uses.

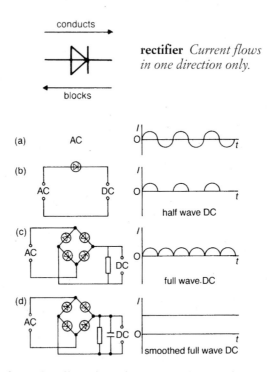

rectifier *Current flows in one direction only.*

rectifier *The effect of rectification on alternating current.*

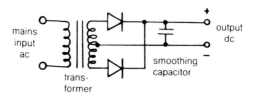

rectifier *A circuit showing a simple rectifier of the type used in power packs for toy trains and in microcomputers.*

red blood cell or **red blood corpuscle** or **erythrocyte** The most numerous **cell** in vertebrate **blood**. Red blood cells are responsible for transporting **oxygen** from the **lungs** to the tissues of the body. In humans, they are biconcave discs and are made in the **bone** marrow. When formed, a red blood cell has a nucleus; however, this is lost by the time the cell enters the blood.

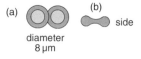

(a)

(b) side

diameter
8 μm

red blood cell *Shape as seen (a) from above and (b) from the side.*

Red blood cells contain haemoglobin. When blood passes through the lungs, haemoglobin combines with oxygen to form the unstable **compound** oxyhaemoglobin. At the tissues this compound breaks down, releasing oxygen to the cells.

A shortage of red blood cells is called anaemia. Compare **white blood cell**.

lungs

haemoglobin + oxygen ⟶ oxyhaemoglobin
tissues

red blood cell *Haemoglobin delivers oxygen to the tissues.*

reduction A substance is reduced if it:

(a) Loses **oxygen**:

$$PbO(s) \; + \; C(s) \; \rightarrow \; Pb(s) \; + \; CO(g)$$
lead oxide   carbon   lead   carbon monoxide

(b) Gains **hydrogen**:

$$Cl_2(g) \; + \; H_2(g) \; \rightarrow \; 2HCl(g)$$
chlorine   hydrogen   hydrogen chloride

(c) Gains **electrons**:

$$Na^+(l) \; + \; e^- \; \rightarrow \; Na(l)$$
sodium ion   electron   sodium

A substance which brings about the reduction of another substance is called a reducing agent. There are several common reducing agents:

Hydrogen	H_2
Carbon	C
Carbon monoxide	CO
Sulphur dioxide	SO_2
Hydrogen sulphide	H_2S

Compare **oxidation**.

refining A process which involves either the removal of impurities from a substance or the extraction of a substance from an impure **mixture**.

reflection One of the three things which can happen to radiation in one medium when it meets the surface of a second medium (see **absorption**, **refraction**). Reflected radiation bounces back into the first medium.

The two laws of reflection allow us to predict where a given *incident ray* will go after reflection at a given point, the *point of incidence*. Angles are measured from the normal. This is the perpendicular to the surface at the point of incidence.

(a) The reflected ray is in the same plane as the incident ray and the normal.

(b) The angle of reflection, r, equals the angle of incidence, i.

reflex action Reflex action occurs in most animals and in vertebrates. It is a rapid response to a **stimulus** over which the animal has no control. Reflex actions can be important in protecting animals from injury; e.g. withdrawing a hand from a hot object.

The *nerve impulse* responsible for reflex actions constitutes a *reflex* arc which is set up when a nerve impulse is initiated in the receptor. The impulse is transmitted along a *sensory neurone* to the **spinal cord** where it

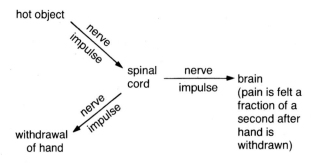

reflex action *Path of nerve impulses.*

crosses a *synapse* to a *motor neurone*. This pathway enables the response to be very rapid.

When a reflex arc operates, nerve impulses are also sent from the spinal cord to the brain. Thus, although the response is initiated by the spinal cord, it is through the brain that the animal is aware of what has happened. *See* **nervous system.**

refraction One of three things which can happen to **radiation** in one medium when it meets the surface of a second (see **absorption, reflection**). After refraction the radiation travels on through the second medium, almost always in a different direction. If the direction is not different, the radiation has not been refracted. The two laws of refraction allow us to predict where a given *incident ray* will go after refraction at a point, the *point of incidence*. Angles are measured from the *normal* or *perpendicular* to the surface at the point of incidence.

(a) The *refracted ray* is in the same plane as the incident ray and the normal.

(b) For a given pair of media, the sine of the angle of incidence (*i*) divided by the sine of the angle of refraction (*r*) is constant (Snell's law).

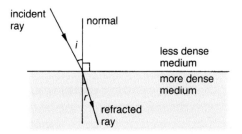

refraction *Refraction of light from a less dense medium to a more dense medium.*

This constant is the *refractive index (n)*. It relates to the **speeds** of the radiation in the two media (see also **colour, lens**). When **light** passes from a less dense medium to a more dense medium it is refracted towards the normal in the more dense medium.

This is caused by the change of speed of the light. When light passes from a more dense medium to a less dense medium the converse is true. When light from a submerged object passes from water to air it is refracted. This causes the object to appear less deep than it really is.

Water waves are refracted when passing from water of one depth to water of a different depth.

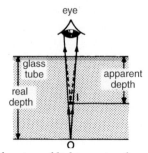

refraction *refraction of light passing from water to air.*

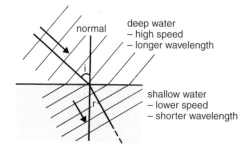

refraction *Refraction of water waves at a change in depth, as a result of the change in speed.*

refrigeration A method for transferring **energy** from a cool box to the warmer outside air. This is the opposite direction to normal net energy transfer and is achieved by a pump powered by a **motor**. The pump circulates a special **fluid** around a closed loop.

When the pump raises the fluid's **pressure** it becomes a **liquid** and gives off **latent heat**. At this stage the liquid goes through the black radiator coil outside so that heat can be transferred to the air. Further around the loop the liquid is forced through the valve to the low pressure side of the loop. The lower pressure causes the liquid to evaporate, absorbing latent heat from the surroundings. At this stage the substance is within the cool box. Devices like this which transfer heat from one place to another are often called *heat pumps*.

relative atomic mass (RAM) The average **mass** of the **atoms** of an **element** compared with an atom of the **isotope carbon-12** which is given the value of exactly 12.

relative molecular mass (RMM) The **mass** of one molecule of a **compound** calculated by adding together the **relative atomic mass** of each **atom** within the molecule. For example:

$$\text{RMM of carbon dioxide (CO}_2) = 12 + (2 \times 16) = 44$$
$$\text{RMM of ammonia (NH}_3) = 14 + (3 \times 1) = 17$$

relay An electrical device which allows a small current to control a large one.

When a current passes through the coil of an **electromagnet** the core attracts the iron armature. As it turns on the pivot it closes the contacts, completing the main circuit. If the coil is switched off, the armature returns to its original position and the main circuit is broken.

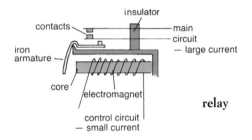

relay

renewable energy sources A general term for sources of energy which are renewed by natural processes. For example, a fast-flowing river is a renewable energy source. The **water** in it can be used to drive **generators** thus producing electricity (**hydroelectricity**). After driving the generators the water eventually flows out to sea. Some of it will evaporate into the air and return onto the land as rain. Thus it will drain into the rivers ready to be used again. Other examples of renewable energy sources are **geothermal aquifers**, **geothermal hot dry rock structures**, **landfill gas**, **solar cells**, **wave generators** and **wind turbines**.

residual current circuit breaker (RCCB) This is a safety device that can be connected between the mains electricity supply and an electrical appliance, such as a lawn mower or hedge trimmer, to protect the user from electrocution in the event of the electric cable being damaged.

The device compares the amount of **current** flowing in the live wire and in the neutral wire of the electric cable. Should these become different in value, such as may result from the cable being damaged, an electric current is induced in the detector winding. This opens the mains switch thus isolating the lawn mower from the electricity supply.

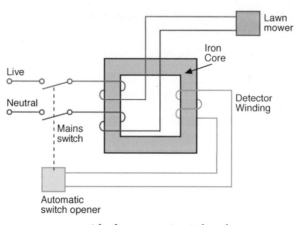

Lawn
mower

Iron
Core

Live

Neutral

Mains
switch

Detector
Winding

Automatic
switch opener

residual current circuit breaker

resistance (R) The opposition of a **circuit** element or section to the flow of **charge** (**current,** I). The unit of resistance is the ohm. The current I in a **metal** sample at a constant **temperature** is proportional to the **voltage,** V across its ends: I *varies with* V (*Ohm's Law*). We define the sample's resistance R from this. $R = V/I$ (rearranging this equation $V = I \times R$, $I = V / R$). All normal circuit components have resistance. As they oppose current there is energy transfer producing a temperature rise (see fuse). The **power,** P, involved is the rate of energy transfer; it is given by:

$$P = V \times I \; (= V^2 / R \; or \; I^2 \times R)$$

Temperature often alters a sample's resistance. In most metals, the increase in temperature makes R rise. In semiconductors, carbon and insulators, R falls. *Resistors* may be arranged in *series* or in *parallel* in a circuit.

When arranged in series the total resistance, R, of the group is given by:

$$R = R_1 + R_2 + R_3$$

When arranged in parallel the total resistance, R, of the group is given by:

$$\frac{1}{R} = \frac{1}{R_1} + \frac{1}{R_2} + \frac{1}{R_3}$$

resonance The large amplitude vibration of a system when it is driven at a **frequency** close to its **natural frequency**. Here are some examples of resonance:
(a) If one sings a note near a piano some of the strings will start to vibrate in sympathy.

(b) The tuning circuit of a radio set will pass high-amplitude signals only if their frequency is near its natural value.

(c) Resonance between the wind and the Tacoma Narrows Bridge (USA) caused the bridge to shake itself apart in 1940.

resources The products which are available within an area. The resources of an area may be described in many different ways. For example:

(a) The amount of food which can be grown.

(b) The mineral deposits which are present.

(c) The **energy** sources it has in the form of **coal**, **natural gas**, **petroleum**, etc.

Many resources are nonrenewable, they can only be used once. For example, coal is a nonrenewable energy source. Once burnt it cannot be used a second time. Much of the **metal** which is obtained from ores is only used once. When the metal article is no longer any use it is thrown away. The continual use of nonrenewable resources results in a resource scarcity.

respiration The chemical reactions by which organisms release **energy** from food such as **glucose**. Respiration may occur in the presence or in the absence of **oxygen**.

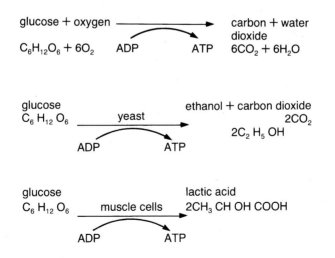

respiration *Anaerobic respiration.*

(a) *Aerobic respiration* occurs in the presence of oxygen within the *mitochondria* of **cells**.

(b) *Anaerobic respiration* occurs in the absence of oxygen within the cytoplasm of cells. It provides less energy than aerobic respiration. **Fermentation** is a form of anaerobic respiration.

reversible reaction A reaction which can go either way depending on the reaction conditions, e.g:

$$Fe_2O_3(s) + 3H_2(g) \rightleftharpoons 2Fe(s) + 3H_2O(g)$$

Hydrogen can react with iron(III) oxide or steam can react with **iron**. In both cases an **equilibrium** will exist with all four substances present unless the products are removed. Reversible reactions are shown in equations by using symbol $\rightleftharpoons$.

rock cycle A process in which existing rocks are continually worn away producing **particles** which are reformed into new rocks. (See illustration opposite).

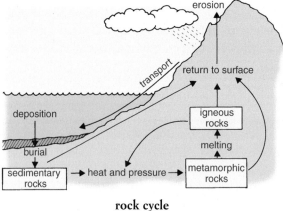

rock cycle

root The part of a flowering plant which normally grows underground in the **soil**. Its functions are:

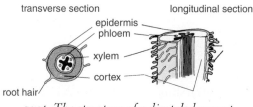

root *The structure of a dicotyledon root.*

(a) To absorb **water and mineral salts** from the soil.

(b) To anchor the plant in the soil.

(c) To act as a food store in some plants, such as the turnip and carrot.

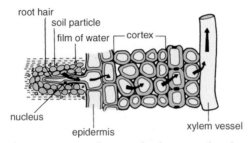

root hairs *Water and mineral salts enter the plant via the root hairs.*

Root hairs project from the cells of the outer layer of the root. These greatly increase the surface area of the root and are where most absorption of water (by **osmosis**) and mineral salts (by **active transport**) take place.

roughage or fibre A component of the human **balanced diet**, consisting mainly of **cellulose** from plant cell walls. Although indigestible to humans, roughage plays an important role in digestion. It adds bulk to the food and enables the **muscles** of the **alimentary canal** to grip the food and keep it moving along by *peristalsis*.

rust The red-brown product of the **corrosion** of **iron** or mild **steel** which has been exposed to water and air. It is hydrated iron(III) oxide ($Fe_2O_2.xH_2O$). Rusting is of great economic importance. Renovation and prevention work on rusting costs British industry, for example, many millions of pounds each year. There are several commonly used methods of preventing rust:

Method	Example
Covering in grease or oil	cycle chain
Covering in paint	motor car body
Covering in plastic	wire mesh fencing
Electroplating	chrome car bumper
Galvanizing	coal bunker
Sacrificial protection	ship's hull

S

saliva Fluid secreted by the salivary glands into the mouths of many animals. Its function is to moisten and lubricate food so that it can be swallowed more easily. In some mammals, including humans, saliva contains the **enzyme** salivary amylase (sometimes called *ptyalin*) which digests **starch** into maltose.

salt This is a **compound** formed when the **hydrogen** of an **acid** is totally or partially replaced by a **metal**. When an acid reacts with a metal the result is a salt and hydrogen gas:

$$Zn(s) + 2HCl(aq) \rightarrow ZnCl_2(aq) + H_2(g)$$

When an acid reacts with a base the result is a salt and **water**:

$$NaOH(aq) + HNO3(aq) \rightarrow NaNO_3(aq) + H_2O(l)$$

Salts may sometimes also be made by the direct combination of two **elements**:

$$2Na(s) + Cl_2(g) \rightarrow 2NaCl(s)$$

If only one hydrogen **atom** of a *dibasic acid* is replaced the result is an *acid salt*, e.g. sodium hydrogen carbonate NaHCO3. The name of the salt is derived from the metal and the acid used:

sulphuric acid → sulphates
nitric acid → nitrates
hydrochloric acid → chlorides

For example, the name of the salt formed by the reaction of copper(II) oxide and dilute sulphuric acid is copper(II) sulphate:

$$CuO(s) + H_2SO_4(aq) \rightarrow CuSO_4(aq) + H_2O(l)$$

secondary sexual characteristics The features which distinguish between adult male and female animals (excluding the reproductive organs). For example, the lion's mane and the antlers on a stag. In humans, secondary sexual characteristics include breast development in females and facial hair in males. The development of these features is usually controlled by **hormones**. Compare **primary sexual characteristics**.

sedimentary rocks Rocks which are formed in layers at the earth's surface. The layers may be formed:

(a) From fragments of rock produced by **weathering** and **erosion**.
(b) By chemical precipitation of dissolved substances.
(c) From the remains of plants and animals.
  Sedimentary rocks are classified in terms of their composition:

Rock	Main minerals present
Limestone	Carbonate minerals calcite and dolomite
Sandstone	Quartz

They may also be classified in terms of their grain size. The following rocks may have similar composition but differ in grain size:

Rock	Diameter of grains in mm
Conglomerate	more than 2
Sandstone	1/16–2
Siltstone	1/256–1/16
Mudstone	less than 1/256

seed This develops from an ovule after **fertilization** in flowering plants. The seeds of a plant are enclosed in a **fruit** (which develops from the **ovary**).
  Within the seed, the **embryo** becomes differentiated into an embryonic shoot bud (plumule) and **root** (radicle) and either one or two seed leaves (cotyledons). Given the right conditions, **germination** will occur and the seed will grow into a new plant. *See* **fruit and seed dispersal**.

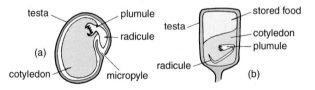

seed *The structure of a seed: (a) broad bean (b) maize grain.*

selective breeding The breeding of animals or plants in order to enhance particular desirable features in their offspring. For example, by breeding only from the cows in a dairy herd which give the highest milk yields, the average milk yield of the whole herd might be expected to increase in time.

sensitivity The ability of plants and animals to respond to **stimuli**, such as **heat**, **light**, **sound**, etc., resulting from changes in their **environment**. Sensitivity makes organisms aware of changes in their environment, thus they can make appropriate responses to any changes which occur. Certain parts of animals, such as the **eyes**, **ears** and **skin**, are specialized in sensing particular environmental stimuli. They are called sense organs or receptors. Similarly, plant tissues such as *shoot tips* are receptors and are important in **tropisms**.

Stimuli from the environment cause responses to be initiated in specialized structures called *effectors*. The **muscles** in our bodies are examples of effectors. The responses which an organism makes constitute its *behaviour*.

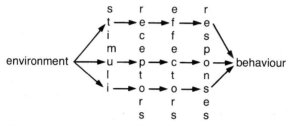

sensitivity *Organisms respond to changes in their environment.*

sexual reproduction A form of reproduction which involves the fusing of two sex cells (**gametes**), one from a male parent and the other from a female parent, to form a zygote. The fusing process is called **fertilization**. The gametes are haploid; however, the resulting zygote has a diploid number of **chromosomes**. After fertilization the zygote divides repeatedly, ultimately resulting in a new organism.

Unlike **asexual reproduction**, the offspring of sexual reproduction are genetically unique (with the exception of identical twins) because they have obtained half of their chromosomes from their male parent and half from their female parent. Thus each fertilization produces a new combination of chromosomes unique to the new organism formed.

shadow A dark shape cast on a surface by an object through which light, a form of **radiation**, cannot pass, as radiation travels in a straight line through a given medium. If the source of radiation is small, and the object large, a sharp shadow is formed.

However, if the source is larger than the object the shadow formed is not sharp and shows two distinct regions. The *umbra*, or full shadow, at the centre, surrounded by the *penumbra*, or partial shadow. No radiation reaches the umbra but some reaches the penumbra. See second diagram below.

See **eclipse**.

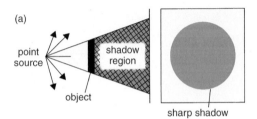

shadow *(a) A point light source casts a sharp umbra; there is no penumbra. See also diagram (b).*

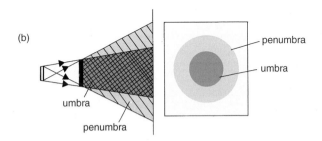

shadow *(b) Shadow produced by a large source of light.*

shoot The part of a plant which is above the soil. This is often composed of the **stem**, **leaves**, buds and **flowers**.

short circuit A path of low **resistance** between two points in a **circuit**. The large flow of **charge** which results from a short circuit draws a heavy **current** from the source. The **heat** produced by this current should be enough to melt the **fuse** and break the circuit before any damage is done to the wiring of the circuit or the load.

SI units The units in which scientific measurements are usually taken (Système International d'Unités). There are seven base units including:

length	metre	m
mass	kilogram	kg
time	second	s

Other units are derived from these base unit:

volume	cubic metres	m^3
density	kilograms per cubic metre	kg/m^3
acceleration	metres per second per second	m/s^2

Some derived units are rather complicated and have been given special names, e.g. for **pressure**; the derived unit is the kilogram metre per second per second per square metre $(kgm/s^2)/m^2$. This is more normally called the **pascal** (Pa).

The size of a unit can be altered by using a series of prefixes:

Prefix	Symbol	Meaning
micro	–	$\times 10^{-6}$ ($\times$ 1/1 000 000)
milli	m	$\times 10^{-3}$ ($\times$ 1/1000)
kilo	k	$\times 10^{3}$ ($\times$ 1000)
mega	M	$\times 10^{6}$ ($\times$ 1 000 000)

For example:

$$1 \text{ millimetre} = 10^{-3} \text{ metre}$$
$$1000 \text{ millimetres} = 1 \text{ metre}$$
$$1 \text{ kilometre} = 1000 \text{ metres}$$
$$0.001 \text{ kilometre} = 1 \text{ metre}$$

skeleton The hard framework of an animal which supports and protects the internal **organs** and gives the animal shape. It also provides a structure for **muscle** attachment and works with muscles to produce movement. In some animals the skeleton lies outside the body and in others the skeleton is contained within the body.

(a)   (b)

exoskeleton *The exoskeletons of (a) an insect and (b) a crustacean.*

(a) *Exoskeleton* (or external skeleton). A skeleton lying outside the body of some invertebrates. Common examples are the tough *cuticle* of insects and the hard shells of crabs. Some organisms have the ability to shed and renew their exoskeletons periodically to allow **growth**. This process is called *moulting* or *ecdysis*.

(b) *Endoskeleton* (or internal skeleton). A skeleton lying within an animal's body. For example, the bony skeleton of vertebrates such as humans. (*See* diagram opposite.) Endoskeletons get bigger as part of the growth process of an animal.

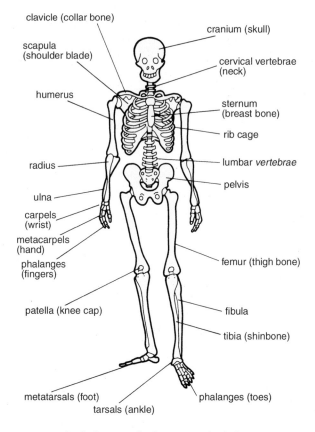

endoskeleton *The human endoskeleton.*

skin A protective layer of **cells**, connective tissue and associated structures that covers most of the body of vertebrates.

The skin of mammals can be divided into two main layers.

(a) The outer layer is the *epidermis*. It consists of:

(i) The *cornified layer*. Dead cells which form a tough protective outer coat.

(ii) The *granular layer*. Living cells which eventually die and form the cornified layer.

(iii) The *Malpigian layer*. Cells which are actively dividing to produce new epidermis.

(b) The inner layer is called the *dermis*. It is a thicker layer than the epidermis and contains **blood capillaries**, hair follicles, sweat glands, receptor cells sensitive to touch, **heat**, cold, pain and **pressure**.

Beneath the dermis there is a layer of **fat** storage cells. These cells act as a food store for the body and also provide heat **insulation**.

The functions of the mammalian skin are:

(a) Protects against injury and the entry of microorganisms which may be harmful to the body.

(b) Reduces water loss by **evaporation**.

(c) Acts as a receptor for certain environmental **stimuli**.

(d) In *homiothermic animals* (animals whose body temperatures remain fairly constant despite environmental conditions) it is important in body **temperature regulation**.

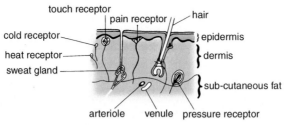

skin *Section through mammalian skin.*

smell The ability of animals to detect odours. In humans the nose is the **organ** of smell. The receptor cells involved are in the nasal cavity. They are sensitive to chemical **stimuli**. *See* **sensitivity**.

sodium A soft grey metallic **element** from **Group I** of the **Periodic Table**. The **metal** is easily cut with a knife to reveal a silvery surface which rapidly tarnishes on exposure to **air**. Sodium reacts vigorously with cold **water**:

$$2Na(s) + 2H_2O(l) \rightarrow 2NaOH(aq) + H^2(g)$$
$$\text{sodium} \quad \text{water} \quad \text{sodium hydroxide} \quad \text{hydrogen}$$

Sodium is stored under oil because of its reactivity to air and water.

Sodium metal is obtained by the **electrolysis** of molten sodium chloride. It is used as a coolant in fast-breeder nuclear reactors and in the manufacture of the petrol additive tetraethyl lead (see **lead**).

Sodium **ions** (Na^+) are an important constituent of the fluids in animal **tissues**. One way in which humans obtain sufficient sodium is to add salt (sodium chloride) to food either during cooking or when eating; however, excessive amounts of sodium cause damage to the body. There are several compounds of sodium which figure prominently in our everyday lives.

(a) *Sodium carbonate* (Na_2CO_3). This compound is made in the Solvay process. It is important in the manufacture of glass. Unlike most carbonates, sodium carbonate does not de compose when heated. Its aqueous solution is alkaline. It may be used to remove permanent hardness from water (see hardness of water).

(b) *Sodium chloride* (NaCl). Known as common salt, and obtained from rock salt. As well as its use for flavouring and preserving foods, it is used in the production of sodium metal and sodium hydroxide.

(c) *Sodium hydrogencarbonate* (NaHCO$_3$). Commonly called baking powder. It is decomposed either by heat or by the action of **acids**.

$$2NaHCO_3(s) \xrightarrow{\text{heat}} Na_2CO_3(s) + CO_2(g) + H_2O$$

sodium   sodium   carbon   water
hydrogen-   cardonate   dioxide
carbonate

$$NaHCO_3(s) + HCl(aq) \rightarrow NaCl(aq) + H_2O(l) + CO_2(g)$$

sodium   **hydrochloric**   sodium   water   **carbon**
hydrogen-   **acid**   chloride   **dioxide**
carbonate

Sodium hydrogencarbonate is also used in fire extinguishers and anti-indigestion powders.

(d) *Sodium hydroxide* (NaOH). This is made by the electrolysis of brine. It is a caustic **alkali** (it will corrode and burn organic material such as flesh) and solutions have **pH>10**. It has many uses in industry, e.g. in the manufacture of soap and paper.

(e) *Sodium nitrate* (NaNO$_3$). This is used as a **fertilizer** and in the preservation of meat.

(f) *Sodium sulphate* (Na$_2$SO$_4$). The hydrated form of this salt (.10H$_2$O) is commonly known as Glauber's Salt. The sulphate is used in the manufacture of paper.

(g) *Sodium thiosulphate* (Na$_2$S$_2$O$_3$). This compound is used in the photographic process. It is used to *fix* the negative and is often called *hypo*. It reacts with unreacted silver bromide, thus ensuring that no further reaction with light occurs.

soil The uppermost layer of the earth's crust. A typical sample of soil contains the following components:

(a) Inorganic **particles** – the result of **weathering** of rocks.
(b) **Water**.
(c) **Humus** – organic material.
(d) **Air**.

(e) **Mineral salts**.

(f) Microorganisms.

(g) Other larger organisms such as earthworms.

Soil is important for several reasons:

(a) It provides a habitat for a wide range of organisms.

(b) It provides plants with a medium for growth and supplies them with water and mineral salts.

(c) The decomposition of dead organisms in soil releases minerals which can be reused by other living organisms.

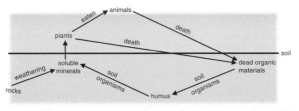

soil *Dead organic material is broken down in the soil.*

Soil is often classified into three distinct types.

(a) *Sandy (light) soil*. This contains a high proportion of larger inorganic particles and hence large air spaces between them. Thus sandy soils are well aerated and have good drainage. Unfortunately good drainage also tends to allow leaching (washing out) of mineral salts.

(b) *Clay (heavy) soil*. This has a high proportion of small particles and hence small air spaces. It retains mineral salts and water but is poorly aerated and may easily become waterlogged.

(c) *Loam soil*. This consists of a balance of particle types and a good humus content. The soil is well aerated and drains well whilst retaining water and mineral salts. Loam is the most fertile soil.

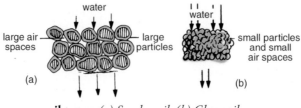

soil types *(a) Sandy soil. (b) Clay soil.*

solar cell A device which converts **energy** from the **Sun** into electrical energy by a photovoltaic process.

solar system This consists of the **Sun**, nine **planets**, some with natural satellites or **moons**, and many asteroids, meteors and comets. The **Earth** is the third nearest planet to the Sun. See Appendix F.

solid A substance whose **atoms** or **molecules** are fixed in positions and do not have the freedom of movement found in a **liquid** or a **gas**. Atoms and molecules are held in a lattice by **bonds**. It is only when these bonds are broken that the atoms and molecules are able to move and the solid *melts*.

solubility The extent to which a solute dissolves in a solvent. In general the solubility of a particular solute increases as the **temperature** of the solvent increases.

In a polar solvent, an **ionic compound** will have a higher solubility than a **covalent compound**. For example, in **water**, sodium chloride is much more soluble than **methane**. Solubility is usually measured in grams of solute per 100 g of solvent at a stated temperature. However other units such as mole/dm^3 and mole/100 g are also used.

sound A form of **radiation** in which **energy** is transferred by means of **pressure waves** in matter, hence sound waves cannot travel through a **vacuum**. A vibrating source pushes particles of matter closer together; as they move apart they move further away from each other than their original positions. This results in alternating high and low pressure regions in the matter. A plot of pressure in the medium at different points at a given time produces a sine wave. A similar pattern is produced by plotting the pressure at a point as time passes.

Sound is thus a wave and shows all wave properties; **absorption**, **reflection**, **refraction**, **diffraction** and *interference*. However sound is a *longitudinal* wave and cannot be polarized.

Like other waves, pressure waves have a **frequency spectrum**. The approximate range of sound which can be detected by the human ear is 20-20 000 Hz. Radiation from the region below 20 Hz is known as *infrasound* and from the region above 20 000 Hz is known as *ultrasound* (see **ultrasonics**). *See* **speed of sound**.

species A unit used in the **classification** of living organisms. It describes any group of organisms which share the same general physical characteristics and can mate and produce fertile offspring. For example dogs, despite a wide variation in shape and size, are all of the same species.

spectrum A type of graph which shows how different types of **radiation** relate to wavelength (and **energy**). For example, **electromagnetic waves** appear on the electromagnetic spectrum. Visible **light** is a part of this spectrum. *See* **colour**.

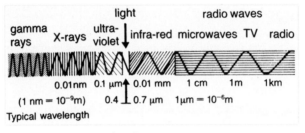

spectrum *The electromagnetic spectrum.*

speed (c) The rate of change of distance. The unit of distance is the metre per second, m/s. Speed is a scalar measure since the direction of movement is not important. This makes it different from velocity which is a *vector* quantity.

speed of light or **velocity of light** The speed of light in empty space, 300 000 000 m/s. In space all **electromagnetic waves** travel at this speed. In matter they move more slowly at speeds which depend on the nature of the substance and the wavelength. *See* **refraction**.

speed of sound The speed of sound travelling through matter depends on the **elasticity** of the medium and its **density**. The density and elasticity of **solids** and **liquids** are far higher than those of **gases**, hence the speed of sound is lowest in gases.

The speed of sound in a particular substance increases with **temperature**. Objects moving through air at a speed greater than sound have *supersonic* speeds and shock waves form around them. These cause Concorde's *sonic boom* and the crack of a long whip.

speed-time graph This shows how the **speed** an object travels varies with time. If the object moves at a constant **acceleration** the graph will be a straight line. A curve indicates that the acceleration of the object is not constant, but varies.

The acceleration of the object is given by the gradient or slope of the graph, y/x. Where acceleration is constant the gradient of the graph is the same at any point. However, where acceleration varies the gradient of the graph changes and must be worked out for each point.

The total distance moved by the object is given by the area under the graph and the time axis. Where acceleration is constant the area of the triangle formed is found using the formula $^1/2$ base × height. However, where acceleration varies the area must be estimated by some means such as counting grid squares.

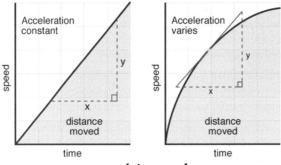

speed-time graph

sperm or **spermatozoon** A microscopic male **gamete** formed in animal **testes**. A sperm usually has a *flagellum* and is able to move. Sperms are released from the male in order to fertilize the female gamete.

spinal cord The part of the central **nervous system** of a vertebrate which is enclosed within and protected by the backbone. It is a cylindrical mass of *nerve cells* which connects the **brain** to the other parts of the body via the spinal nerves. There are three distinct regions in the spinal cord.
(a) An inner layer of grey matter which consists of nerve cell bodies.
(b) An outer layer of white matter which consists of nerve fibres running the length of the cord.
(c) A fluid-filled central canal.

   The spinal cord conducts *nerve impulses* to and from the brain and is also involved in **reflex actions**.

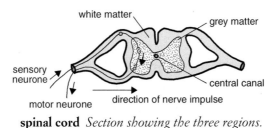

spinal cord *Section showing the three regions.*

spleen An **organ** in the abdomen. In most vertebrates it is found near the **stomach**. It has several important functions:
(a) It produces **white blood cells**.
(b) It destroys worn out **red blood cells**.
(c) It filters foreign bodies from the **blood**.

spore A (usually) microscopic reproductive unit consisting of one or several **cells** which have become detached from the parent organism and will ultimately become a new individual. Spores are involved in both **asexual** and **sexual reproduction** (as **gametes**). They are produced by certain plants, **fungi**, **bacteria** and protozoa. Some spores form a resistant resting stage during a life cycle while others allow a rapid colonization of new **habitats**.

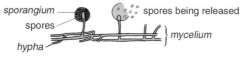

sporangium spores being released
spores
hypha mycelium

spore *Spore release in the bread mould Mucor.*

stamen The male part of the **flower**. Each stamen consists of a stalk (filament) and at the end of each filament there is an anther. Within the anthers there are **pollen** grains containing the male **gametes**.

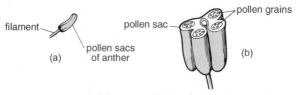

filament pollen sac pollen grains
pollen sacs
(a) of anther (b)

stamen *(a) Stamen. (b) Anther cut open.*

star A heavenly body, like the **Sun**, which produces its own **light**. The brightness of a star is indicated by its *magnitude*: the smaller the magnitude, the brighter the star. The source of the **energy** emitted by stars is nuclear **fusion**.

starch A *polysaccharide* **carbohydrate** which consists of chains of **glucose** units. Starch is important as an **energy** store in plants, and is synthesized by plants during **photosynthesis**. It is readily converted into glucose by *amylase* **enzymes**.

states of matter Substances can exist in three states of matter; **solid**, **liquid** and **gas**. The state in which a substance exists depends upon its

temperature and the **pressure** exerted on it. Not all substances exist in all states. The initial letters (s) = solid, (l) = liquid, (g) = gas, (and (aq) = aqueous solution) are sometimes used in chemical **equations** to show the states of the reactants and products:

$$2Na(s) + 2H_2O(l) \rightarrow 2NaOH(aq) + H_2(g)$$

static electricity Stationary electric charges, as opposed to moving ones, i.e. current. Static often appears on an insulator when it is rubbed and charge may last for a long time. The discharge of static which has built up on a gramophone record can be heard as a 'crackle'. The discharge of static is a potential hazard in areas such as mines, factories and aircraft fuelling systems. A spark in the presence of inflammable material in the air may result in an explosion.

steel Steels are **alloys** which contain **iron** as the main constituent. Other elements present will determine the properties of the steel. The two best known types of steel are *mild steel* and *stainless steel*.
(a) Mild steel contains iron with small amounts of **carbon**. It is used for car bodies and household goods such as freezers and cookers. It is cheap to make, but it **rusts** easily in the presence of moisture and oxygen and so it must be protected. Common methods of protection are painting, greasing, enamelling, galvanizing and coating in plastic.
(b) Stainless steel contains iron, carbon, chromium and nickel. Common uses of stainless steel are cutlery, surgical instruments and sink tops. It is not corroded by oxygen, however its use is limited because it is much more expensive to produce than mild steel.
The iron made in a **blast furnace** contains impurities which make it brittle, e.g. carbon (more than is needed for mild steel). This iron is made into steel by blowing oxygen through the molten iron and thus oxidizing the impurities which are given off as **gases**. When this is done other **elements** are added to the melt to produce the steel required.

stem The part of a flowering plant which bears the buds, **leaves** and **flowers**. The functions of the stem are:
(a) To transport **water**, **mineral salts** and **carbohydrate**.
(b) To raise the leaves above the soil so they get the maximum amount of **air** and **light**.
(c) To raise flowers, thus aiding the process of **pollination**.
(d) In green stems, **photosynthesis**.

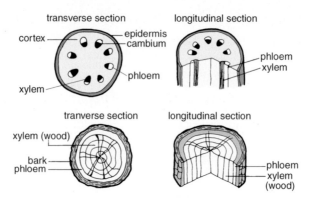

stem *Structure of the stem of a dicotyledonous plant. Transverse and longitudinal sections of (a) a young stem and (b) an older stem.*

stimulus Any change in for example **temperature** or **pressure** in the **environment** of an organism which may produce a response in the organism. *See* **sensitivity**.

stomach A muscular sac in the anterior region of the **alimentary canal**. In vertebrates food passes into the stomach from the mouth via the oesophagus.

In the stomach, food is mechanically churned by the peristaltic action of the muscular walls and the **digestion** of **protein** is started. In herbivores the stomach has several chambers, for cellulose digestion.

The amount of time which food spends in the stomach depends on its nature. From the stomach it passes into the small intestine through a ring of muscle called the *pyloric sphincter*.

structural formula The formula of a substance which shows the **bonds** between its **atoms** and their positions relative to each other.

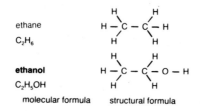

structural formula *The structural formula gives more information than the molecular formula.*

sucrose A **carbohydrate** belonging to the group of **sugars** called *disaccharides*. Sucrose is the white crystalline substance which we use at home and commonly call sugar. It is obtained from sugar beet or sugar cane.

sugar A common name for a series of sweet **compounds** which are either monosaccharides or disaccharides. For example:
(a) monosaccharides – **glucose**, fructose, ribose.
(b) disaccharides – maltose, **sucrose**.
 See **carbohydrate**.

sulphur (S) A yellow nonmetallic **element** which may exist in two forms or *allotropes*. *Rhombic sulphur* is the stable form below 96 °C and *monoclinic sulphur* is stable above that **temperature**.
   Sulphur belongs to **Group** VI of the **Periodic Table** and is reactive to both **metals** and **nonmetals**. It occurs in nature both uncombined and as metal sulphides such as galena (PbS) and pyrite (FeS). Sulphur and sulphur-containing compounds are also found in **petroleum, natural gas** and **coal**. Such impurities are a problem as the **combustion** of any **fuel** containing them will produce sulphur dioxide, one of the main causes of **acid rain**. In modern processes sulphur and sulphur-containing compounds are either removed from fuels before combustion, or any sulphur dioxide gas produced is removed from the exhaust gases. Sulphur is an important constituent of some drugs, such as *sulphonamides*, and is used to vulcanize rubber; however, its main use is in the manufacture of **sulphuric acid** in the contact process.

sulphuric acid (H_2SO_4) A very important chemical used in many processes. It is a colourless, oily **liquid** which is a strong **acid** and a vigorous oxidizing agent (see **oxidation**). It is made from *sulphur dioxide* in the *contact process*. Sulphuric acid reacts chemically in several ways:

(a) As an acid. Dilute sulphuric acid reacts with **metals, bases** and
    carbonates to form *sulphates*:

$$Mg(s) + H_2SO_4(aq) \rightarrow MgSO_4(aq) + H_2(g)$$
$$CuO(s) + H_2SO_4(aq) \rightarrow CuSO_4(aq) + H_2O(l)$$
$$ZnCO_3(s) + H_2SO_4(aq) \rightarrow ZnSO_4(aq) + H_2O(l) + CO_2(g)$$

   Concentrated sulphuric acid reacts with chlorides and nitrates to
   form hydrogen chloride and nitric acid respectively:

$$H_2SO_4(l) + NaCl \rightarrow NaHSO_4(s) + HCl(g)$$
$$H_2SO_4(l) + NaNO_3 \rightarrow NaHSO_4(s) + HNO_3(g)$$

(b) As a dehydrating agent. Concentrated sulphuric acid is sometimes used to dry gases. It can also be used to remove the **atoms** which make **water** from substances.

(c) As an *oxidizing agent*. Although **copper** cannot directly replace **hydrogen** from acids, the metal is oxidized by concentrated sulphuric acid:

$$Cu(s) + 2H_2SO_4(l) \rightarrow CuSO_4(s) + SO_2(g) + 2H_2O(l)$$

       anhydrous
       copper(II)
       sulphate

The reaction between concentrated sulphuric acid and water is a very **exothermic reaction**. When diluting concentrated sulphuric acid, it is important always to add the acid to the water (by stirring) and NOT the other way around. (See page 262).

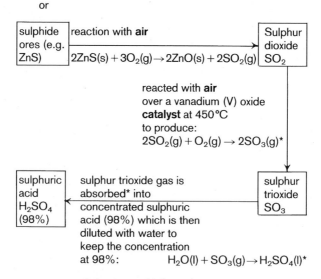

sulphuric acid *The contact process.*

Sun The **star** at the centre of the **solar system**. The Sun is the source of most of the **energy** available on **Earth** either directly, as **light** and **heat**, or indirectly, as the source of energy for plant **growth**. The Earth's axis is permanently tilted at an angle of $66.5°$ to the plane of its orbit. The tilting of the axis is responsible for the following:

(a) The changes in the altitude of the midday sun throughout the year. The sun is highest at midday on June 21st (*summer solstice*) and lowest at midday on December 22nd (*winter solstice*).

(b) The variation in length of the days (and nights) throughout the year. The longest day is June 21st and the shortest December 22nd.

(c) The four seasons.

In the summer in Britain the weather is generally warmer. This is the result of the sun being higher in the sky (the sun's rays are more intense) and the longer day length. In the winter the weather is generally colder. The sun's rays are less intense and day length is shorter.

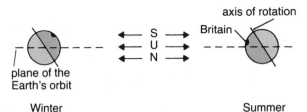

axis of rotation

S
U
N

Britain

plane of the
Earth's orbit

Winter

Summer

sun *The tilting of the earth's axis is responsible for the difference in the seasons.*

T

taste The ability of animals to detect flavours. In humans the receptor cells for taste are called taste buds. They are sensitive to certain chemical **stimuli**. They are restricted to the mouth, particularly on the tongue. There are four types of taste bud sensitive to sweet, sour, salt and bitter. Taste and **smell** often work together to identify different foods. *See* **sensitivity**.

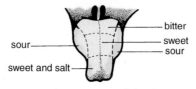

taste *A taste-map of the tongue.*

tectonic theory This theory considers the crust of the **Earth** to be made of huge rigid blocks called plates. There are seven major plates together with several minor ones.

The plates are thought to slide on molten **magma**. Plates meet along *plate boundaries* and the areas near to the plate boundaries are called *plate margins*. The process of forming new geological structures, as a result of the movement of plates, is called *plate tectonics*. *See* **earthquake**, **volcano**.

teeth Structures within the mouths of vertebrates which are used for biting, tearing and crushing food before it is swallowed.

Enamel. This covers the exposed surface of the tooth (the crown). It contains calcium phosphate and is the hardest substance in a vertebrate's body. It is well suited to biting food without itself being damaged.

Dentine. This substance is similar to **bone** and forms the inner part of the tooth.

Pulp. The soft **tissue** in the centre of the tooth. It contains the **blood capillaries**, which supply food and **oxygen**, and the nerve fibres which register pain if the tooth is damaged.

Root. The part of the tooth within the gum and embedded in the jawbone. The outer surface of the root is covered in a substance called cement. A series of fibres hold the tooth in place. One end of them is embedded in the cement and the other in the jawbone.

Most human beings are *omnivores*. The adult human jaw contains four types of teeth.

Incisors. Chisel-shaped teeth at the front of the mouth used for biting off pieces of food.

Canines. Pointed teeth at each side of the incisors. They are used for ripping off pieces of food. In wild animals which are carnivores, such as lions and tigers, these teeth are often very large and also used for killing prey. In herbivores these teeth are often small or missing completely.

Premolars. Grinding teeth found between the canines and the molars.

Molars. Together with premolars these teeth are often called *cheek teeth* because of the position they occupy at the back of the mouth. They are broad-crowned teeth which crush and grind food before it is swallowed. They are found in both herbivores and omnivores but in carnivores they are replaced by *carnassial* teeth.

See **dental formula**.

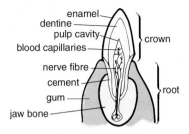

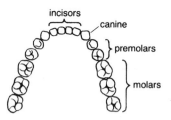

teeth *The arrangement of teeth in an adult human jaw (bottom and top jaws are the same).*

teeth *Structure of a tooth.*

temperature (T) The **heat** level of a substance. It relates to the mean **energy** of the **particles** of the sample. There are several temperature scales used to show the degree of temperature. In daily life we use *Celsius* (*centigrade*) or *Fahrenheit*. In science the absolute scale is also commonly used.

Temperature is measured using a **thermometer**.

Scale	Unit	Ice	Steam
Celsius	degree C, °C	0 °C	100 °C
Fahrenheit	degree F, °F	32 °F	212 °F
Kelvin	kelvin, K	273 K	373 K

temperature regulation *Homoiothermic* animals maintain their body temperature within a narrow range despite the temperature of their **environment**. For example, the temperature of a healthy human being is always around 37 °C. This is essential so that the normal reactions of **metabolism** can take place. Here are some ways in which birds and mammals regulate their temperature:

(a) Fat under the skin (subcutaneous) acts as an insulator.

(b) Hair in mammals and feathers in birds trap air which is a good insulator.

(c) In mammals the **evaporation** of sweat from the surface of the **skin** has a cooling effect.

(d) Blood vessels near the surface of the skin constrict in response to cold. This diverts blood away from the skin surface so less heat is lost (*vasoconstriction*).

(e) Blood vessels near the surface of the skin dilate in response to heat. This brings blood up to the skin surface so more heat is lost to the atmosphere (*vasodilation*).

testis The **sperm**-producing reproductive **organ** of male animals. Male vertebrates such as men have a pair of testes which, in addition to sperms, also produce **hormones**.

thermometer A device used to measure **temperature** or 'hotness'. This relates to the mean **energy** of the **particles** of the sample. Thermometers work by measuring something which varies with temperature, e.g.:

(a) A column of **liquid** in a glass tube; as the liquid gets hotter it expands and the level rises up the tube.

(b) The **voltage** between two **metals**.

Other properties, such as the **pressure** of a **gas**, the **resistance** of a wire, or the colour of a hot surface, may also be used. In each case the device is used at certain known temperatures (often the **boiling and freezing points** of **water**) and is then marked in degrees according to the scale used.

thermoplastic A **polymer** which softens when heated and can be moulded and remoulded into new shapes. **Nylon** and *polychloroethene* are examples of thermoplastics. Compare **thermosetting plastic**.

thermosetting plastic A **polymer** which cannot be softened and reshaped by heating. The application of heat to a thermosetting plastic results in its decomposition. *Bakelite* and *formica* are examples of thermosetting plastics. Compare **thermoplastic**.

thermostat A device which is used to keep the **temperature** in a place within a particular range. Thermostats are present in a number of common household devices such as cookers, refrigerators, irons, freezers and heating boilers. Many thermostats use a **bimetallic strip**.

three-pin plug The standard method of joining a cable to a mains socket. A mains cable has three wires. These are colour-coded as shown in the table below. Each wire goes to a particular pin in the plug.

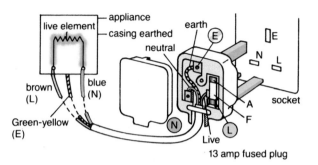

Three-pin plug

13 amp fused plug

The size of the fuse fitted in the plug depends on the **power** of the device to which it is connected.

Wire	Colour of plastic coating	Purpose
live	brown	carries **current**
neutral	blue	returns current
earth	green/yellow	safety

tidal barrage A wall built across a river estuary. The wall contains *turbogenerators* which transfer **kinetic energy** from the **tides** into **electricity**. There are no tidal barrages currently operating in Britain but studies have been carried out on several river estuaries (including the Mersey and Severn) with a view to building tidal barrages in the future. *See* **renewable energy sources.**

tide The alternating rise and fall of the sea which occurs about every twelve hours. Tides are due to the gravitational pull of the **Moon** and, to a lesser extent, the gravitational pull of the **Sun.**

Spring tides occur when the Moon, Earth and Sun are in a straight line. The combined pull of the Moon and Sun result in very high tides and equally low tides. The difference in the level of the sea at high tide and low tide is very large.

Neap tides occur when the pull of the Sun is at right angles to that of the Moon. The difference in the level of the sea at high tide and low tide is quite small.

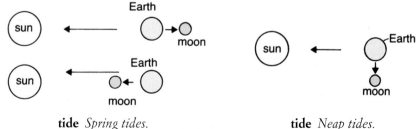

tide *Spring tides.*   **tide** *Neap tides.*

tissue In **multicellular** organisms tissue is a group of similar **cells** which are specialized to carry out a specific function within the organism. For example, **muscle** in animals and xylem in plants. The human body contains many different types of tissue.

transformer A device normally used to transfer electrical **energy** with a change in **voltage**.

An **alternating current** input to the primary coil causes an alternating magnetic field in the core of the transformer. In turn, this changing field induces an alternating current at the secondary (output) coil. The ratio of the output voltage to the input voltage equals the ratio of the number of turns of wire in the output and input coils, i.e.:

$$\frac{V_2}{V_1} = \frac{N_2}{N_1}$$

For a *step down* transformer, $V_2 < V_1$ hence $N_2 < N_1$, and for a *step up* transformer the reverse applies $V_2 > V_1$ hence $N_2 > N_1$. The electrical power in the coil of a transformer is calculated by multiplying the current passing through the coil by the voltage across it. The **efficiency** of a transformer is often measured as the ratio of useful output power to total input power.

$$\text{efficiency} = \frac{\text{useful output power}}{\text{total input power}}$$

No energy transfer device has an efficiency of 100%; however, many transformers are very close to this. *See* **electromagnetic induction**.

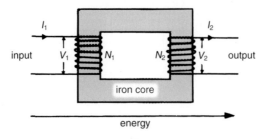

transformer

transistor A semiconductor device whose output **current** depends on signals to the base. Its main uses are in amplifiers and oscillators.

transpiration The **evaporation** of water vapour from plant leaves via tiny pores called *stomata*. The rate at which transpiration occurs depends on several environmental factors.

(a) **Temperature** – increased temperature increases the rate at which water evaporates and thus increases transpiration.
(b) Humidity (water content of air) – increased humidity causes the atmosphere to become saturated with water vapour, thus reducing transpiration.
(c) Wind – increased air movements increase the rate at which water evaporates by pre venting the atmosphere immediately around the stomata from becoming saturated; thus it increases the rate of transpiration.

It follows from the above that the transpiration rate will be greatest in warm, dry, windy conditions. If the rate of water loss by transpiration exceeds the rate of water uptake through the **roots** the plant may begin to wilt.

tropisms In plant **growth**, movement in response to a **stimulus** such as **light**. These movements are related to the direction of the stimulus, the plant organ growing either towards or away from it. Tropisms are named by adding a prefix which refers to the stimulus involved:
(a) Geotropism – a response to gravity.
(b) Phototropism – a response to light.
(c) Chemotropism – a response to chemicals.
(d) Hydrotropism – a response to **water**.

Tropisms can be either positive or negative depending on whether the response is in the same direction or the opposite direction to the stimulus. They are important because they allow plants to grow in such a way that they can get the maximum benefit from their **environment** in terms of water, light, etc.

Tropisms are caused by a plant **hormone** or *auxin* which accelerates growth by stimulating cell division and elongation. Uneven distribution of auxins causes uneven growth which eventually leads to bending.

U

ultrasound Sound waves of higher **frequency** than can be detected by the human ear (about 20 kHz). Some other animals are able to hear some of the ultrasonic range. For example, the *sonar* system of bats typically works at 50 kHz or higher. Ultrasound of megahertz frequencies has many uses in modern life. For example, detecting explosive mines at sea, checking on babies in the womb (typical value 2.25 MHz), and cleaning engine parts.

ultraviolet A region in the **spectrum** of **electromagnetic waves**. The approximate wavelength range is 10^{-10}–10^{-7} m and the approximate **frequency** range is 10^{15}–10^{18} Hz. Ultraviolet **radiation** is produced by white-hot objects and certain **gas** discharges. It affects photographic film and causes fluorescence and photo-electric effects. Human **skin** makes **vitamin D** with this radiation and light skins are tanned by it. However, over-exposure is harmful to both the **eyes** and skin. The **ozone layer** limits the amount of ultraviolet radiation which reaches the earth's surface.

unicellular (of an organism) Consisting of only one **cell**. Unicellular organisms include protozoans, **bacteria** and some algae. Compare **multicellular**.

Universe The total content of all that exists. The *Steady State Theory* and the *Big Bang Theory* both offer an explanation why the Universe is as we see it today.

The steady state theory, first proposed in 1948, suggests that the Universe was not created but has always existed and is thus infinitely old. The various stellar events that are witnessed from the **Earth** indicate the Universe is in a state of continuous change and this will continue to be so forever. This theory is unable to explain important observations made by modern astronomers and, since the mid-1960s, has been superseded in popularity by the big bang theory.

The big bang theory suggests that all of the matter and **energy** which now exists in the Universe was once concentrated into a small volume of unimaginable **density**. Some fifteen thousand million years ago, there was a massive explosion throwing matter and energy out in all directions creating the Universe. The existence of cosmic background radiation, and the continuous expansion of the Universe provide evidence that supports this theory.

uranium This **element** is a **metal** which has 92 **protons**. Its **atoms** are the most massive of all natural elements. Uranium has three **isotopes**; uranium-234, uranium-235 and uranium-238. All three are radioactive, being sources of **alpha particles**.

The **nuclei** of uranium-235 atoms can undergo nuclear **fission** when they absorb **neutrons**. The process of nuclear fission itself produces more neutrons. Under certain conditions, a **chain reaction** may take place. This is the basis of **nuclear power** and the atomic bomb.

urea The main nitrogenous excretory product of mammals. Urea is made in the **liver** as a product of the *deamination* of excess **amino acids**: It travels from the liver to the **kidney** via the bloodstream and is excreted in the **urine**.

$$H_2N-C-NH_2 \quad \text{urea}$$
$$\overset{\|}{O}$$

urine An aqueous solution of **urea** and salts produced by the **kidneys** of mammals. It is stored in the bladder before being discharged via the urethra.

uterus or womb A muscular cavity found in most female mammals. The uterus contains the **embryo** during its development. It receives **ova** from the oviduct and connects to the exterior of the body via the vagina.

V

vaccine A mixture containing dead or inactivated microorganisms normally responsible for a disease. It is given to stimulate (see **stimulus**) an *immune response* in a person from which he or she gains immunity to the disease. Vaccines are not quick-acting but rely on the gradual build up of **antibodies** in the person's **blood**. *See* **immunization**.

vacuum A space from which the **air** has been removed, thus containing very few **particles** of matter. A cubic metre of air at standard pressure contains around 10^{25} particles; a good vacuum on **Earth** contains about 10^{14}. Between the Earth and the **Moon** the value is far lower, and between galaxies there may be a true vacuum (containing no particles). Nevertheless the vacuum of outer space contains **energy** in forms other than matter such as **force** fields, **radiation** and very low **mass** particles called neutrinos.

valency The number of **bonds** which an **atom** forms with other atoms. More precisely, the valency of an **element** is the number of **electrons** that it needs to form a **compound** or radical. The electrons may be lost, gained or shared with another atom.

Some elements always have the same valency. **Sodium** always has a valency of 1. It gives one electron away when it forms the sodium **ion** Na^+ in compounds like sodium chloride NaCl. **Oxygen** always has a valency of 2. It accepts two electrons when it forms the oxide ion O^{2-} or it forms **covalent compounds** such as CO_2 and SO_2. *Transition elements* have more than one valency, e.g. cobalt = 2 or 3; **copper** =1 or 2; **iron** = 2 or 3. The table below shows the valencies of some common elements and the ions they form.

+1		+2		+3	
Lithium	Li^+	Calcium	Ca^{2+}	Aluminium	Al^{3+}
Sodium	Na^+	Magnesium	Mg^{2+}	Iron(III)	Fe^{3+}
Potassium	K^+	Zinc	Zn^{2+}		
Silver	Ag^+	Iron(II)	Fe^{2+}		
Ammonium	NH_4^+	Lead	Pb^{2+}		

	−1		**−2**		**−3**
Fluoride	F⁻	Sulphide	S^{2-}	Phosphate	PO_4^{3-}
Chloride	Cl⁻	Oxide	O^{2-}		
Bromide	Br⁻	Carbonate	CO_3^{2-}		
Iodide	I⁻	Sulphate	SO_4^{2-}		

valency

variation The difference in characteristics between two members of the same species. There are two main types of variation.

(a) *Discrete variation.* In this type of variation only particular values within a given range are possible. For example, a person's **blood** may belong to group A, B, AB or O but not somewhere between two groups, i.e. there are no possible intermediate forms. Discrete data does not show a *normal distribution.*

(b) *Continuous variation.* In this type of variation the variable may take any value within a particular range. For example, if a person is between 1.90 and 1.91 m tall they may be 1.900, 1.901, 1.902 m etc. The only limitation is the accuracy with which we measure the height. Another example in humans is **mass**. Continuous variation throughout a population shows a normal distribution about an average value called the mean. (We often represent continuous data as if it were discrete data by placing it in groups.)

Variation within a species results from either inherited or environmental factors or a combination of both. For example, a human being inherits genes influencing height but is also subject to environmental factors such as nutrition. Inherited variations are considered to be the basis of **evolution** by **natural selection**.

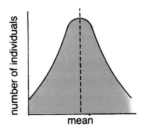

variation *A continuous variation, showing the mean.*

vegetative reproduction **or propagation** Asexual reproduction
in plants by an outgrowth of some kind from the parent plant.
The outgrowth may occur in several different ways. The table shows
some examples:

Form of outgrowth	Example of plant
bulb	daffodil, onion
corm	crocus, gladiolus
rhizome	iris
stolon	strawberry
tuber	potato, dahlia

vein A **blood** vessel which transports blood from the **tissue** of the body to
the **heart**. In mammals, veins carry deoxygenated blood, with the
exception of the pulmonary vein which carries oxygenated blood from the
lungs to the heart. Veins are formed from groups. of smaller blood vessels
called *venules* which carry blood from **capillaries**. The blood pressure in
veins is less than in **arteries**, and this accounts for the differences in their
structures. Veins have much thinner walls than arteries and have valves to
prevent blood flowing away from the heart.

The largest veins in humans are the *superior vena cava*, which carries
blood from the head, neck and upper arms to the heart, and the *inferior
vena cava* which carries blood from the rest of the body to the heart.

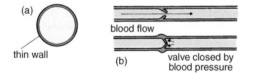

vein (a) Section through a vein. (b) Valve operation in veins.

velocity (v) An object's displacement in unit time. The unit used is the
metre per second, m/s. Velocity is a vector (compare **speed**). To find
velocity we divide the displacement, s, by the time taken, t; $v = s/t$. The
slope at any point on a graph of displacement against time gives the
velocity at that instant.

vertebral column **or backbone** A series of interlocking bones (called
vertebrae) and/or *cartilages* which runs along the back from the skull to the
tail in vertebrates. It is the principal longitudinal supporting structure of
the body. It also protects the **spinal cord** which runs down its centre.

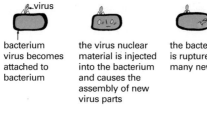

virus *A virus infecting a bacterium.*

virus Viruses are the smallest known living organisms, having diameters between 0.025 and 0.25 microns. They are **parasites** infecting animals, plants and **bacteria**. A virus particle consists of a **protein** coat surrounding a length of nucleic acid, either **DNA** or RNA.

Virus infections in humans include measles, polio and influenza.

vitamins These are organic **compounds** which are required in small amounts by living organisms. Like **enzymes**, vitamins play a vital role in chemical reactions within the body and often regulate an enzyme's action. If the human diet contains insufficient amounts of vitamins this will result in deficiency diseases.

Vitamin	Rich sources	Effects of deficiency
Vitamin A	milk, liver, butter, fresh vegetables	night blindness, retarded growth
Vitamin B_1	yeast, liver	Beri-beri: loss of appetite and weakness
Vitamin B_2	yeast, milk	pellagra: skin infections, weakness, mental illness
Vitamin C	citrus fruits, fresh green vegetables	scurvy: bleeding gums, loose teeth, weakness
Vitamin D	eggs, cod liver oil	rickets: abnormal bone formation
Vitamin E	fresh green vegetables, milk	thought to affect reproductive ability
Vitamin K	fresh vegetables	blood clotting impaired

vitamins *The properties of some important vitamins.*

volcano An opening in the **Earth**'s crust through which **magma** comes to the surface of the earth as lava. The lava leaves the volcano through an opening called a vent. Ash and gases, including water vapour, are also given out into the air.

Volcanoes may be active, dormant or extinct. The effects of a volcanic eruption may be considerable. In AD 79, the eruption of the volcano Vesuvius buried the nearby town of Pompeii in hot ash to a depth of 3 m. When the volcano Krakatoa erupted in 1883 there was an explosion which was heard 5000 km away and around 18 km³ of material was thrown into the air. *See* **earthquake**.

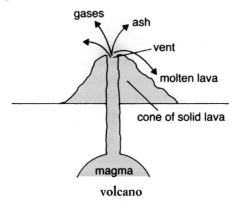

volcano

voltage or potential difference   The electrical **pressure** which drives electric charges round a **circuit**. The unit is the volt, V. **Electricity** will only flow between two places which have a potential difference between them. One volt is the potential difference between two points which allows 1 **joule** of **energy** to be obtained when 1 *coulomb* of **charge** moves between the points.

$$\text{volts} = \text{joules/coulombs}$$

volume   The amount of space taken up by an object. The **SI unit** of volume is the cubic metre, m³.

The volumes of **solids** and **liquids** are fairly constant. The volume of a **gas** depends greatly on the **temperature** and **pressure**.

volume
of cuboid = a × b × c

volume
of cylinder = πr^2 h

volume
of sphere = $\frac{4}{3}\pi r^3$

volume
of cone = $\frac{1}{3}\pi r^2$ h

volume

n the **Earth** and
e atmosphere.
in, snow, etc.)
enters the rocks
ns via lakes and
r their own use.
ocesses such as
hrough
ciers and in the

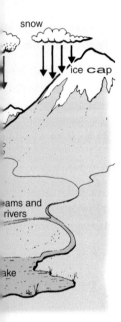

$2H_2O(l)$

ids on the **Earth**, and is
ess **liquid**. Here are some of its

    $0 \ °C$
    $100 \ °C$
    $1.0 \ g/cm^3$
  °C)

on solidification, thus ice is less
bes burst when the water in
t conduct electricity; however, it
l (H_2SO_4) or **alkali** (NaOH) are
e **hydrogen** and **oxygen**:

$+ O_2(g)$

e of water by adding an
cobalt chloride, which will

$CuSO_4.5H_2O(s)$
    blue
$CoCl_2.6H_2O(s)$
    pink

measure its boiling point.
found in the **atmosphere**, rivers
ater is continually flowing from
lants and animals; this is

solvent. Water has polar bonds
m chloride (NaCl), as well as

oltage (volts)

rticles). Most
l. Transverse
irection of **energy**

ys contains small amounts of
and depending on the source of
ls. Some dissolved solids make

water cycle The continuous movement of water betwe
the **atmosphere**. Water evaporates from the oceans into t
From here it falls on the earth (in several forms, such as r
and also freezes out of the air as frost and ice. This water
and soil and will eventually find its way back into the oce
rivers. Plants and animals take water out of the ground fo
This water will eventually return to the atmosphere, by p
transpiration, **respiration** and sweating, or to the oceans
excretion. Large quantities of water are stored as ice in g
polar ice caps.

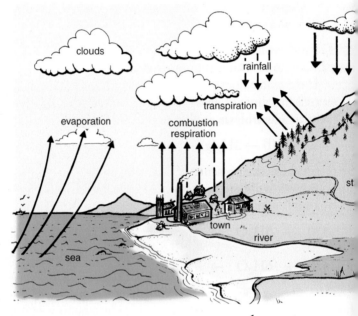

water cycle

watt The unit of **power**. This is a measure of the rate at
done.

(a) In mechanics: $\text{power (watts)} = \dfrac{\text{work done (\textbf{joules})}}{\text{time taken (seconds)}}$

(b) In electricity: power (watts) = **current** (amperes) × v

wave A type of **radiation** (the other type consisting of p
waves can be described as either transverse or longitudin
waves consist of *vibrations* whose direction is across the

transfer. With longitudinal waves the vibration direction is the same as the direction of energy transfer.

The distance taken up by a single cycle of a wave is called the *wavelength λ*).

The wavelength (λ), frequency (*f*) and speed (*c*) of a wave are related in the basic wave equation:

$$\text{speed} = \text{frequency} \times \text{wavelength} \ (c = f \times \lambda)$$

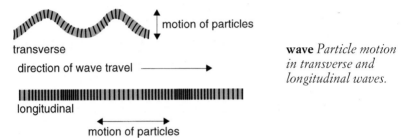

transverse

motion of particles

direction of wave travel

longitudinal

motion of particles

wave *Particle motion in transverse and longitudinal waves.*

All waves can show **absorption**, **reflection**, **refraction**, **diffraction** and interference. *See* **electromagnetic waves, sound, water.**

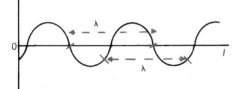

wave *Wavelength.*

wave generators Devices which transfer the **kinetic energy** of the waves at sea into **electricity**. The device is sited on the shore line and the up and down movement of the water level, caused by the waves, is used to force air through a turbine which drives a **generator**.

See **renewable energy sources**.

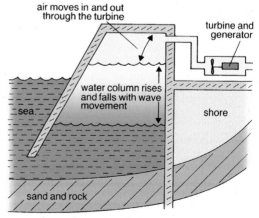

air moves in and out through the turbine

turbine and generator

water column rises and falls with wave movement

sea

shore

sand and rock

wave generator

weather The condition of the **atmosphere**. The factors which make up the weather include rain, **temperature**, humidity, wind speed and direction and atmospheric pressure.

weathering The processes (physical or chemical) by which rock is fragmented or altered. Physical processes which break rock up include:
(a) **Expansion** and contraction caused by daily **temperature** changes.
(b) The freezing and expansion of water in cracks.
Chemical weathering processes include:
(a) The dissolving of minerals in water.
(b) The reaction of minerals with **oxygen** from the **air** or with carbonic acid (formed when **carbon dioxide** dissolves in **water**).
See **erosion**.

weight (W) The **force** which an object exerts downwards because of gravitational attraction. Like all forces, it is measured in newtons (N). The weight of an object depends on its **mass**, m, and the strength of the gravitational field, g; $W= mg$. This gives, as a unit for g, the newton per kilogram N/kg. An object's mass is thought to be constant. The value of g depends on where the object is, hence the weight of the object also depends on where it is. On the **Earth** (g =10 m/s^2 or 10 N/kg), the weight of a 5 kg object is $5 \times 10 = 50$ N. On the **Moon** (g = 1.6 m/s^2 or 1.6 N/kg), its weight would be 5x1.6=8N. In a place, such as outer space, where there is very little gravitational force (i.e. g= 0) acting on an object its weight will be zero. *See* **gravity**.

white blood cell, white blood corpuscle or leucocyte One of the types of **blood cell** found in most vertebrates. Their function is to defend the body against microorganism infection. This is achieved in two ways:
(a) Some white blood cells produce **antibodies** which react with invading microorganisms and render them harmless.
(b) Some white blood cells engulf and digest invading microorganisms. This process is called *phagocytosis*.
Compare **red blood cell**.

wind turbine A device in which **kinetic energy** from the wind is transferred into **electricity**. A huge propeller, attached to a shaft, is driven around by the wind in much the same way as the sails of a windmill. The rotating shaft is used to drive a **generator**. Wind turbines are devices which make use of **renewable energy sources**. They are often grouped together in *wind farms*.

X – Y

X-rays A region of the **spectrum** of **electromagnetic waves**. The approximate wavelength range is 10^{-12}–10^{-10} m and the approximate **frequency** range is 10^{18}–10^{21} Hz. X-rays are produced by firing **electrons** at **metals**. It is a very penetrating **radiation** and will easily pass through flesh, but is stopped by **bone** and other dense materials. X-rays are widely used in medicine to take pictures of bones. A photographic plate is placed behind the area of the body to be investigated and the area of the body is then exposed to an X-ray source. The bones show up as light areas on the photographic plate.

yeast A general name for a group of microorganisms which are very useful to humans. Yeasts contain **enzymes** which convert **sugars** into **ethanol**. This process is called **fermentation**.

Appendix A:
Abbreviations and Symbols

A list of some useful common abbreviations and symbols you may encounter in scientific literature.

A	mass number; also Ampère – unit of electric current
aq	state symbol for aqueous solution usually as (aq)
A_r	relative atomic mass
atm	atmosphere – a unit of pressure
a	alpha Greek letter
b	beta Greek letter
b.p.	boiling point
C	Celsius as in °C degree Celsius; also Coulomb –unit of electric charge (quantity of electricity)
cm^3	cubic centimetre, unit of volume
DC	direct current – the type of electricity produced from a battery
dm^3	cubic decimetre 1 litre, unit of volume
E	symbol for emf of a cell
e or e$^-$	electron
emf	electromotive force
g	gram – unit of mass; state symbol for gas usually as (g) ; also acceleration due to gravity
H	enthalpy (ΔH =enthalpy change)
I	electric current
J	Joule – unit of energy
k	prefix meaning 'one thousand times' i.e. kg = 1000 g
K	Kelvin– unit of temperature, $1K \equiv 1$ °C
l	state symbol for liquid usually as (l)
m	mass; also metre – unit of length
M	molar unit of concentration (molarity) e.g. 2M
m^3	cubic metre unit of volume
ml	millilitre, 1/1000 of 1 litre $\equiv 1cm^3$
mol.	mole unit of amount of substance
m.p.	melting point
M_r	relative molecular mass n neutron
N	Newton – unit of force
nm	nanometer – one thousand-millionth of a metre
NTP	Normal temperature and pressure
p	proton; also pressure
Pa	Pascal – unit of pressure
p.d.	potential difference
pH	relates to a scale of acidity, e.g. pH = 1 very strongly acidic
Q	electric charge, quantity of electricity
s	state symbol for solid usually as (s); also second unit of time
STP	standard temperature and pressure
t	time
T	temperature
$T_{1/2}$ or $t_{1/2}$	half life (of radioactive isotope)
UV	ultraviolet
V	volume; also electrical potential difference (p.d.); also volt – unit of p.d.
Z	atomic number

Appendix B:
Circuit Symbols

A list of the standard symbols and names of the main circuit elements you are likely to meet.

Conductors

—— conductor

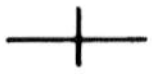

crossing conductors

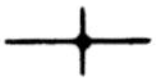

conductor junction

sliding contact

simple switch

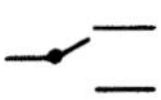

two-way switch

Sources

general supply to output

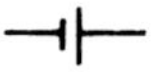

cell (short arm is negative)

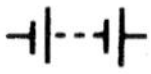

battery of cells

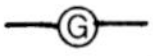

generator

aerial

Passive elements

resistor or resistive element

fuse

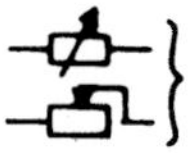

} variable resistor

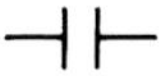

capacitor

electrolytic capacitor (black arm is cathode)

inductor

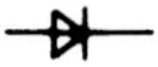

diode (allows current to the right)

signal lamp

filament lamp

earth

Electromagnet elements

—Ⓐ—   ammeter

—Ⓥ—   voltmeter

=◁   speaker

    microphone

—Ⓜ—   motor

    transformer

Vacuum devices

    diode

    cathode ray tube

Semiconductor devices

    pn junction (diode)

light-emitting diode

photo-diode

pnp transistor

npn transistor

light sensitive pn diode

Appendix C:

Characteristics of living things

For an organism to be considered as 'living' it must demonstrate *all* of the following features:

Movement	The ability to change position either of all, or, of part of the body.
Excretion	The ability to remove from the body waste materials produced by the organism.
Respiration	The ability to release energy by the breakdown of complex chemicals.
Reproduction	The ability to produce offspring.
Irritability	The ability to sense and respond to changes in the environment.
Nutrition	The ability to take in or manufacture food that can be used when required as a source of energy or as building materials.
Growth	The ability to increase in size and complexity through the production of new cell material.

Appendix D:

The Differences between
Plants and Animals

Plants	Animals
Cell surrounded by cellulose cell wall	No cellulose cell wall
Large vacuoles in cells	Vacuoles when present filled with cell sap only small
Large cells with definite shape	Small irregularly shaped cells
Only restricted movement possible	Free movement possible
Response to stimulus slow	Rapid response to stimulus
Cells contain chloroplasts (chlorophyll)	No chloroplasts (chlorophyll)
Photosynthesise	Must obtain food from external sources

These characteristic differences should only be regarded as guidelines. Attempts to classify certain organisms within these terms of reference will be difficult and has provided scientists with great problems, for example, bacteria, fungi, viruses.

Appendix E:

The Major Groups of Living Organisms

THE ANIMAL KINGDOM (major phyla)

(a) **INVERTEBRATES** Animals without a vertebral column (backbone)

Phylum Protozoa Microscopic **unicellular** animals.

Phylum Porifera Porous animals often occurring in colonies, for example, sponges.

Phylum Coelenterata Tentacle-bearing animals with stinging cells.

Phylum Platyhelminthes Flatworms.

Phylum Annelida Segmented worms.

Phylum Molluscs Soft-bodied animals often with shells.

Phylum Arthropoda Jointed limbs; exoskeleton.

Phylum Echinodermata Spiny-skinned marine animals.

(b) **VERTEBRATES** (Phylum Chordata) Animals with a vertebrate column.

Class Mammalia (mammals) Hair; constant temperature; young suckled with milk.

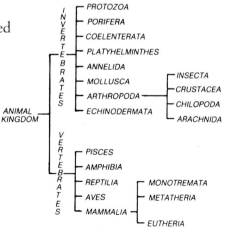

Summary

THE PLANT KINGDOM (major phyla)

Phylum Thallophyta Unicellular and simple **multicellular plants**.

Class Algae Photosynthetic; includes unicellular, filamentous, and multicellular types.

Class Fungi Heterotrophic; including both parasites and saprophytes.

Phylum Bryophyta Green plants with simple leaves and showing alternation of generations; moist habitats.

Class Hepaticae (liverworts)

Class Musci (mosses)

Phylum Pteridophyta (ferns; bracken; horsetails) Green plants, with roots, stems, leaves, and showing alternation of generations.

Phylum Spermatophyta Seed producing plants.

Class Gymnospermae Seeds produced in cones.

Class Angiospermae Flowering plants; seeds enclosed within fruits.

Bacteria and viruses do not meet the criteria necessary to be placed in either the animal or plant kingdoms.

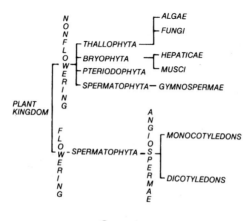

Summary

Appendix F:

The Solar System

Planet	Diameter at equator in km	Mean distance from the sun in million km	Time taken to orbit the sun in years
Mercury	4 800	57.6	0.24
Venus	12 320	107.5	0.62
Earth	12 680	148.8	1.00
Mars	6 720	226.4	1.88
Jupiter	141 900	773.3	11.86
Saturn	120 160	1417.8	29.57
Uranus	46 880	2852.8	84.75
Neptune	50 400	4468.8	164.79
Pluto	5 900	5865.6	248.43